Aouss Gabash

Flexible Optimal Operations of Energy Supply Networks

Aouss Gabash

Flexible Optimal Operations of Energy Supply Networks

With Renewable Energy Generation and Battery Storage

Südwestdeutscher Verlag für Hochschulschriften

Impressum / Imprint
Bibliografische Information der Deutschen Nationalbibliothek: Die Deutsche Nationalbibliothek verzeichnet diese Publikation in der Deutschen Nationalbibliografie; detaillierte bibliografische Daten sind im Internet über http://dnb.d-nb.de abrufbar.
Alle in diesem Buch genannten Marken und Produktnamen unterliegen warenzeichen-, marken- oder patentrechtlichem Schutz bzw. sind Warenzeichen oder eingetragene Warenzeichen der jeweiligen Inhaber. Die Wiedergabe von Marken, Produktnamen, Gebrauchsnamen, Handelsnamen, Warenbezeichnungen u.s.w. in diesem Werk berechtigt auch ohne besondere Kennzeichnung nicht zu der Annahme, dass solche Namen im Sinne der Warenzeichen- und Markenschutzgesetzgebung als frei zu betrachten wären und daher von jedermann benutzt werden dürften.

Bibliographic information published by the Deutsche Nationalbibliothek: The Deutsche Nationalbibliothek lists this publication in the Deutsche Nationalbibliografie; detailed bibliographic data are available in the Internet at http://dnb.d-nb.de.
Any brand names and product names mentioned in this book are subject to trademark, brand or patent protection and are trademarks or registered trademarks of their respective holders. The use of brand names, product names, common names, trade names, product descriptions etc. even without a particular marking in this works is in no way to be construed to mean that such names may be regarded as unrestricted in respect of trademark and brand protection legislation and could thus be used by anyone.

Coverbild / Cover image: www.ingimage.com

Verlag / Publisher:
Südwestdeutscher Verlag für Hochschulschriften
ist ein Imprint der / is a trademark of
OmniScriptum GmbH & Co. KG
Heinrich-Böcking-Str. 6-8, 66121 Saarbrücken, Deutschland / Germany
Email: info@svh-verlag.de

Herstellung: siehe letzte Seite /
Printed at: see last page
ISBN: 978-3-8381-3838-1

Zugl. / Approved by: Ilmenau, TU, Diss., 2013

Copyright © 2014 OmniScriptum GmbH & Co. KG
Alle Rechte vorbehalten. / All rights reserved. Saarbrücken 2014

This book summarizes my works during my PhD research at the Ilmenau University of Technology from 01.07.2009 to 01.07.2013. I would like to thank my wife *Olaa* and my children *Abdullah* and *Abdulrahman* for their love and support during producing this book.

Ilmenau, 01.04.2014

Aouss Gabash

Abstract

Due to environmental and fuel cost concerns more and more wind- and solar-based distributed generation (DG) units are embedded in distribution networks (DNs). It is, however, a well-known fact that renewable energy generators are highly fluctuating sources, and therefore, energy storage systems such as battery storage systems (BSSs) are considered as a solution to handle such fluctuations. In general, DG units and/or BSSs convert traditional passive DNs (PDNs) into active DNs (ADNs). Consequently, it is important to investigate the impact and benefits of integrating such entities in conventional DNs.

This dissertation presents a systematic study consisting of modeling, simulation, and optimization of dynamic operations of energy supply networks with embedded renewable generation and storage. Based on complex power flow models, different optimization problems are mathematically formulated and solved.

In this work, novel mathematical models and a new combined problem formulation for active-reactive optimal power flow (A-R-OPF) in PDNs (without DG units and BSSs) and ADNs (with DG units and BSSs) are studied. Typically, DNs consist of two different networks in terms of voltage levels, namely, low-voltage and medium-voltage DNs. For this reason, investigations are carried out separately on both networks. Modeling procedures for PDNs, ADNs, and energy prices are presented. These procedures serve as the basis for this work. Then, simulation studies in PDNs are made to analyze its operating characteristics. In particular, the operation of on-load-tap-changers of main transformers is highlighted. Moreover, an optimization framework is introduced to minimize the total energy losses in PDNs.

In ADNs, two voltage levels with two real case studies are separately considered. On the low-voltage level, a high penetration level of photovoltaic (PV) systems (PVSs) is considered in the network in order to reveal the impact of such a scenario. In particular, the reactive power capability of the inverters of these PVSs is explored. The total revenue from the installed PVSs is maximized whilst the total cost of energy losses and demand is minimized. Using different price models many interesting results are found, e.g., no need to use BSSs in low-voltage DNs for accommodating expected spilled PV energy. On the medium-voltage level, a DN with a high penetration of wind energy and BSSs is considered. In this case, the total revenue from wind parks and BSSs is maximized and the total cost of energy losses is minimized. It is found that a huge reduction in energy losses and reactive energy imports can be achieved. To prolong the life of BSSs only one fixed charge/discharge cycle every day is considered. The solution provides an optimal operation strategy which ensures the feasibility and enhances the revenue significantly. However, due to the fact that the profiles of renewable energy generation, demand and prices vary from day to day a fixed operation of BSSs cannot be optimal.

A flexible battery management system is proposed to adapt to such variations. This is accomplished by optimizing the lengths (hours) of charge and discharge periods of BSSs for each day, leading to a complex mixed-integer nonlinear program (MINLP). An iterative two-stage framework is proposed to address this problem. In the upper stage, the integer variables (i.e., hours of charge and discharge periods) are optimized and delivered to the lower stage. In the lower stage the A-R-OPF problem is solved by a NLP solver and the resulting objective function value is brought to the upper stage for the next iteration. This procedure will converge when number of iterations is reached. Using this flexible system a considerably higher revenue can be achieved.

Zusammenfassung

Bedingt durch Umweltbelange und steigende Kosten für fossile Brennstoffe werden immer mehr Wind- und Solaranlagen (distributed generation, DG) in Verteilernetzen (distribution networks, DNs) installiert. Es ist eine bekannte Tatsache, dass die Einspeisung durch erneuerbare Energieträger starken Schwankungen unterliegt. Ein möglicher Lösungsansatz zur Behandlung dieser Schwankungen ist die Nutzung von Energiespeichersystemen wie z. B. Batteriespeichersysteme (BSS). Der Einsatz solcher Systeme verwandelt traditionelle passive Verteilernetze (PDNs) in aktive Verteilernetze (ADNs). Folglich ist es wichtig, die Auswirkungen und Vorteile der Integration solcher Einheiten in konventionelle Verteilernetze zu untersuchen.

In dieser Dissertation wird eine systematische Untersuchung (bestehend aus Modellierung, Simulation und Optimierung) des dynamischen Betriebs von Energieversorgungsnetzen mit eingebetteten erneuerbaren Energieträgern und Speichersystemen vorgenommen. Basierend auf komplexen Lastflussmodellen werden verschiedene Optimierungsprobleme mathematisch formuliert und gelöst.

In dieser Arbeit werden neue mathematische Modelle und eine neue Problemformulierung für den kombinierten optimalen Lastfluss von Wirk- und Blindleistung (active-reactive optimal power flow, A-R-OPF) in PDNs (ohne DG-Anlagen und BSSs) und ADNs (mit DG-Anlagen und BSSs) vorgestellt. Typischerweise enthalten DNs zwei Spannungsebenen, nämlich Nieder- und Mittelspannung. Deshalb werden Untersuchungen in den beiden Spannungsebenen getrennt durchgeführt. Modellierungsverfahren für PDNs, ADNs und Energiepreise werden vorgestellt. Diese Verfahren dienen als Grundlage der vorliegenden Arbeit. Darauf aufbauend werden

Simulationsstudien in PDNs zur Analyse der Betriebseigenschaften durchgeführt. Insbesondere wird die Auswirkung des Betriebs der Laststufenschalter der Haupttransformatoren hervorgehoben. Darüber hinaus wird ein Optimierungsverfahren zur Minimierung der gesamten Energieverluste in PDNs vorgestellt.

In ADNs werden zwei Spannungsebenen mit jeweils zugehörigen realen Fallstudien getrennt betrachtet. Auf der Niederspannungsebene wird eine hohe Einspeisungsrate von PV-Anlagen (photovoltaic systems, PVSs) angenommen, um die Auswirkungen eines solchen Szenarios zu zeigen. Insbesondere wird die Fähigkeit der Inverter dieser PV-Anlagen zur Erzeugung von Blindleistung untersucht. Die Gesamteinnahmen aus den installierten PV-Anlagen werden maximiert, während gleichzeitig die Gesamtkosten der Energieverluste und die Nachfrage minimiert werden. Durch die Verwendung unterschiedlicher Preismodelle können viele interessante Ergebnisse generiert werden, z. B. besteht keine Notwendigkeit, BSSs in Niederspannungsnetzen für die Aufnahme überschüssiger PV-Energie zu installieren. Auf der Mittelspannungsebene wird ein DN mit einer hohen Einspeisungsrate von Windenergie und BSSs betrachtet. In diesem Fall werden die Gesamteinnahmen der Windparks und BSSs maximiert, während die Gesamtkosten der Energieverluste minimiert werden. Es zeigt sich, dass eine enorme Reduktion der Energieverluste und der Blindleistungsimporte erreicht werden kann. Um die Lebensdauer der BSSs zu verlängern wird nur ein fester Lade-/ Entlade-Zyklus pro Tag betrachtet. Diese Lösung liefert eine optimale Betriebsstrategie, welche die Zulässigkeit gewährleistet und den Profit signifikant erhöht. Aufgrund der Tatsache, dass die Profile der erneuerbaren Energien, der Nachfrage und der Preise von Tag zu Tag variieren, ist ein feststehender Betrieb der BSSs allerdings nicht optimal.

Weiterhin wird ein flexibles Batterie-Management-System zur Behandlung solcher Schwankungen vorgestellt. Dies wird durch die Optimierung der Lade- und Entladezeiten der BSSs für jeden Tag erreicht. Daraus resultiert ein komplexes gemischt-ganzzahliges nichtlineares Optimierungsproblem (mixed-integer nonlinear program, MINLP). Für dessen Lösung wird ein iteratives zweistufiges Verfahren eingeführt. In der oberen Stufe werden die ganzzahligen Variablen (d. h. Lade- und Entladezeiten) optimiert und an die untere Stufe weitergegeben. In der unteren Stufe wird das A-R-OPF Problem mit einem NLP-Löser gelöst und der resultierende Wert der Zielfunktion wird an die obere Stufe für die nächste Iteration weitergegeben. Dieses Verfahren konvergiert, wenn eine Anzahl von Iterationen erreicht ist. Die Verwendung dieses flexiblen Ansatzes resultiert in bedeutend höheren Profiten.

Contents

NOMENCLATURE XI

1 INTRODUCTION 1

    1.1 MOTIVATION 1
    1.2 CONTRIBUTIONS AND DISSERTATION STRUCTURE 4
    1.3 SOFTWARE TOOLS FOR SIMULATION AND OPTIMIZATION 7

2 LITERATURE REVIEW 9

    2.1 DIRECT AND ALTERNATING CURRENT 9
    2.2 MATHEMATICAL FORMULATION OF OPTIMAL POWER FLOW 12
    2.3 OPF WITH RENEWABLE ENERGIES AND STORAGE SYSTEMS 14
        2.3.1   *OPF without DG Units* *15*
        2.3.2   *OPF with DG Units* *16*
        2.3.3   *Energy Storage for Power Systems* *17*
        2.3.4   *OPF with DG Units and BSSs* *19*
        2.3.5   *Traditional Electricity Market* *20*
        2.3.6   *Electricity Market with DG Units and BSSs* *21*
    2.4 FLEXIBLE A-R-OPF WITH RENEWABLE ENERGIES AND BSSS 22

3 MODELING PROCEDURES 25

    3.1 BACKGROUND 25
    3.2 MODELING OF PASSIVE DISTRIBUTION NETWORKS 27
        3.2.1   *Load Model and Bus Types* *28*
        3.2.2   *Feeder Model of Distribution Networks* *28*
        3.2.3   *On-Load Tap Changer Transformer* *29*
        3.2.4   *Power Flow in PDNs* *30*
        3.2.5   *Newton-Raphson Method* *30*
    3.3 MODELING OF ACTIVE DISTRIBUTION NETWORKS 33
        3.3.1   *Wind Power* *33*
        3.3.2   *Photovoltaic Power* *34*
        3.3.3   *Battery Storage* *37*
        3.3.4   *Power Flow in ADNs with Battery Storage* *41*
        3.3.5   *Definition of Infinite Bus* *43*
    3.4 MODELING OF ENERGY PRICES 44
        3.4.1   *Forward Active-Reactive Energy Prices* *44*
        3.4.2   *Feed-in-Tariffs and Reverse Active Power Flow* *45*
        3.4.3   *Charge-Remuneration Rates for Battery Storage* *45*
        3.4.4   *Meter-Based Method for Charging and Remunerating* *47*

4 SIMULATION AND OPTIMIZATION IN PASSIVE DISTRIBUTION NETWORKS 49

4.1 SIMULATION IN PDNs 49
    4.1.1 Dynamic Power Flow in PDNs 49
    4.1.2 Control System of an OLTC Transformer 50
    4.1.3 Case Studies 50
4.2 OPF IN PDNs UTILIZING OLTCs CAPABILITY 64
    4.2.1 Optimal Voltage Regulation in PDNs 64
    4.2.2 Proposed Method 67
    4.2.3 A Case Study 68

5 ACTIVE-REACTIVE OPTIMAL POWER FLOW IN ACTIVE DISTRIBUTION NETWORKS 71

5.1 A-R-OPF FOR LOW-VOLTAGE ADNs 71
    5.1.1 Modeling of Network Demand, Generation and Energy Prices 71
    5.1.2 A-R-OPF Utilizing PV-DG Reactive Power Capability 72
    5.1.3 A Case study 76
    5.1.4 Conclusions 80
5.2 A-R-OPF FOR MEDIUM-VOLTAGE ADNs 83
    5.2.1 Modeling of Network Demand, Generation and Energy Prices 83
    5.2.2 A-R-OPF with Wind-Battery Stations 83
    5.2.3 A Case study 88
    5.2.4 Conclusions 95

6 FLEXIBLE OPTIMAL OPERATION OF BATTERY STORAGE SYSTEMS FOR ENERGY SUPPLY NETWORKS 97

6.1 PROBLEM DESCRIPTION 97
    6.1.1 Varying Demand, Generation and Energy Prices Profiles 98
    6.1.2 Operational Constraints of BSSs 100
    6.1.3 Market Strategies 102
6.2 PROBLEM FORMULATION AND SOLUTION FRAMEWORK 103
    6.2.1 Problem Formulation 103
    6.2.2 A Two-Stage Solution Framework 104
    6.2.3 A Search Method for the Upper Stage Problem 105
6.3 A CASE STUDY 109
6.4 CONCLUSIONS 116

7 SUMMARY AND FUTURE RESEARCH ASPECTS 123

BIBLIOGRAPHY 137

APPENDIX A: IEEE-RTS LOAD DATA   127
APPENDIX B: DATA FOR THE LOW-VOLTAGE NETWORK   128
APPENDIX C: DATA FOR THE MEDUIM-VOLTAGE NETWORK   130
APPENDIX D: SOFTWARE IMPLEMENTATION OF DSI-1   132
APPENDIX E: SOFTWARE IMPLEMENTATION OF DSI-2   135
BIBLIOGRAPHY   137

Nomenclature

Functions

f	Function
F	Vector of f
F	Total objective function value
$F_{w.b}$	Total revenue from wind power and BSSs
F_{loss}	Total cost of energy losses
F_{pv}	Total revenue from PV power
F_{demand}	Total cost of demand
Losses	Total energy losses
ΔP	Active power mismatches
ΔQ	Reactive power mismatches

Parameters

$B(i,j)$	Imaginary component of the complex admittance matrix elements
B_l	Line capacitive susceptance
$C_{pr.p.art}(h)$	Average remuneration tariff for active energy during hour h
$C_{pr.p}^{(A)}$	Two-tariff price model of active energy
$C_{pr.p}^{(B)}$	Three-tariff price model of active energy
$C_{pr.p}^{(C)}$	24-hour-tariff price model of active energy
$C_{pr.p}(h)$	Active energy price during hour h
$C_{pr.q}(h)$	Reactive energy price during hour h
D	Annual operation days
$E_{BSS}(i)$	Installed capacity of BSS i

Nomenclature

$E_{max}(i)$	Upper bound of energy in BSS i
$E_{min}(i)$	Lower bound of energy in BSS i
$G(i,j)$	Real component of the complex admittance matrix elements
G_l	Line conductance
it_{max}	Maximum number of iterations
Ipv	Set of PVSs
Iw	Set of wind parks
Ib	Set of BSSs
N	Total number of buses
N_{line}	Total number of lines
$N_1:N_2$	The nominal turn-ratio of an OLTC
n	Total number of charge/discharge cycles per day
$P_{peak}(i)$	Daily peak active power demand at bus i
$P_d(i,h)$	Active power demand at bus i during hour h
PT_{peak}	Active power transformer peak load
$PT(h)$	Active power transformer load during hour h
$P_g(i,h)$	Active power of a CGE at bus i during hour h
$P_{pv}(i,h)$	Active power of PV generation at bus i during hour h
$P_r(i,h)$	Active power of RGE at bus i during hour h
$P_w(i,h)$	Active power of wind generation at bus i during hour h
$P_{PV}(i)$	Rated power of PVS i
$P_W(i)$	Rated power of wind park i
p	Total number of charge/discharge cycles in the lifetime
$Q_d(i,h)$	Reactive power demand at bus i during hour h
$Q_g(i,h)$	Reactive power of a CGE at bus i during hour h
QT_{peak}	Reactive power transformer peak load

$QT(h)$	Reactive power transformer load during hour h
r	Replacement period in years
R_l	Line resistance
R_t	Transformer nominal resistance
$S_{S1.max}$	Upper bound of apparent power at slack bus
$S_{PCS.max.b}(i)$	Upper bound of apparent power of BSS i
$S_{PCS.max.pv}(i)$	Upper bound of apparent power of PVS i
$S_{l.max}(i,j)$	Upper bound of apparent power of line between bus i and j
T_1, T_3	Daily time intervals for low price
T_2	Daily time interval for high price
T_{final}	Time horizon
$T_{initial}$	Initial time in a time horizon
t_{min}	Minimum bound of integer control variables (daily time intervals)
t_{max}	Maximum bound of integer control variables (daily time intervals)
u_{min}	Minimum bound of continuous control variables
u_{max}	Maximum bound of continuous control variables
$V_{min}(i)$	Lower bound of voltage at bus i
$V_{max}(i)$	Upper bound of voltage at bus i
$V_{S0}(h)$	Primary voltage of a TR during hour h
$V_{S1.ref}(h)$	Voltage reference of a TR during hour h
v	Wind speed (m/s)
v_{ci}	Cut-in speed of wind turbine (m/s)
v_{co}	Cut-off speed of wind turbine (m/s)
v_r	Rated speed of the wind turbine (m/s)

Nomenclature

X_l	Line inductive reactance
x_{min}	Minimum bound of state variables
x_{max}	Maximum bound of state variables
X_t	Transformer nominal leakage reactance
Z_t	The equivalent transformer impedance
η_{ch}	Battery charge efficiency
η_{dis}	Battery discharge efficiency
ε	Tolerance
$\alpha_{P1.fw}$	Upper bound of active power in forward direction at slack bus
$\alpha_{Q1.fw}$	Upper bound of reactive power in forward direction at slack bus
$\alpha_{P1.rev}$	Upper bound of active power in reverse direction at slack bus
$\alpha_{Q1.rev}$	Upper bound of reactive power in reverse direction at slack bus

State variables

a:1	The nominal tap-ratio of the autotransformer
$E(i,h)$	Energy level in BSS i during hour h
it	Iteration number
$P_{S1}(h)$	Active power injected at slack bus during hour h
$P(i,h)$	Active power injection at bus i during hour h
$Q_{S1}(h)$	Reactive power injected at slack bus during hour h
$Q(i,h)$	Reactive power injection at bus i during hour h
$Q_{disp.ava.b}(i,h)$	Available reactive power of BSS i during hour h
$Q_{disp.ava.pv}(i,h)$	Available reactive power of PVS i during hour h
$S(i,j,h)$	Apparent power flow from bus i to bus j during hour h
$S_{PCS.b}(i,h)$	Apparent power of BSS i during hour h

$S_{PCS.pv}(i,h)$	Apparent power of PVS i during hour h
$V_e(i,h)$	Real component of complex voltage at bus i during hour h
$V_f(i,h)$	Imaginary component of complex voltage at bus i during hour h
$V(i,h)$	Voltage at bus i during hour h
V	Voltage amplitudes
x	State variable
X	Vector of x
ΔX	Correction vector of X
θ	Phase angles

Control variables

$P_{ch}(i,h)$	Active power charge of BSS i during hour h
$P_{dis}(i,h)$	Active power discharge of BSS i during hour h
$Q_{disp.b}(i,h)$	Reactive power dispatch of BSS i during hour h
$Q_{disp.pv}(i,h)$	Reactive power dispatch of PVS i during hour h
$Q_{disp.r}(i,h)$	Reactive power dispatch of REG i during hour h
t	Integer variable
t_1, t_3	Integer variables for time periods of charge
t_2	Integer variable for time period of discharge
u	Continues variable
$V_{S1.ref}(h)$	Discrete variable for voltage amplitude during hour h
$\beta_{c.pv}(i,h)$	Curtailment factor of PV power at PVS i during hour h
$\beta_{c.r}(i,h)$	Curtailment factor of REG power at REG i during hour h
$\beta_{c.w}(i,h)$	Curtailment factor of wind power at wind park i during hour h

Nomenclature

Acronyms

AC	Alternating current
AC-OPF	AC optimal power flow
A-R-OPF	Active-reactive optimal power flow
ADN	Active distribution network
B-PCS	Battery-PCS
BSS	Battery storage system
CISO	California Independent System Operator
CGE	Conventional generator
CT	Current transformer
DC	Direct current
DC-OPF	DC optimal power flow
DEM	Demand
DG	Distributed generation
DN	Distribution network
DoD	Depth of discharge
DOP	Dynamic optimizer
DS	Distribution system
DSO	Distribution system operator
DSI	Dynamic simulator
EEG	Erneuerbare-Energien-Gesetz
e.m.f.	Electromotive force
ESS	Energy storage system
FACTS	Flexible alternating current transmission systems
FBMS	Flexible battery management system
FIT	Feed-in-tariff

g		Equality constraints
G		Generating plant
GA		Genetic algorithm
GAMS		General algebraic modeling system
GEGEA		Green Energy and Green Economy Act
GTO		Gate turn-off thyristor
HV		High-voltage
HV-TN		High-voltage transmission network
IESO		Independent electricity system operator
ISO		Independent system operator
ISO-NE		Independent System Operator New England
J		Jacobian
LMP		Locational marginal price
LP		Linear programming
LV		Low-voltage
M		Meter
MCP		Market clearing price
MCC		Marginal Cost of Congestion
MCL		Marginal Cost of Losses
MINLP		Mixed-integer nonlinear programming
MV		Medium-voltage
MV-DN		Medium-voltage distribution network
NLP		Nonlinear programming
NE		Total number of equations
N.V.		Number of variables
OLTC		On-load-tap-changer

OPF	Optimal power flow
OTS	One-at-a-time-search
PCS	Power conditioning system
PDN	Passive distribution network
PF	Power factor
PSB	Polysulfied-Bromine
pu	Per unit
PV	Photovoltaic
PV-PCS	Photovoltaic-PCS
PVS	Photovoltaic system
REA	Renewable Energy Act
REG	Renewable energy generator
S_0	Primary side of a TR
S_1	Secondary side of a TR
SMEC	System Marginal Energy Cost
StrEG	Stromeinspeisungsgesetz
SVC	Static VAR Compensator
TN	Transmission network
TOU	Time of use
TR	Transformer
TS	Transmission system
VRB	Vanadium redox battery
VHV	Very high-voltage
VT	Voltage transformer
ZMCP	Zonal MCP

1 Introduction

1.1 Motivation

Renewable energies, such as wind and solar, are being more and more considered as alternative solutions to cover the increasing demand for energy and minimize CO_2 emissions. It is, however, a well-known fact that renewable energy generators (REGs) are very fluctuating energy sources and the already existing power system and its electrical networks may not be able to accommodate a large scale of such kind of generators. Briefly, a large amount of renewable energies can be lost due to many technical and economical constraints. Therefore, many studies and researches have been made to reveal the impact and benefits of such kinds of generation units. To effectively increase the penetration of renewable energies many solutions have been recently proposed such as upgrading the already existing electrical networks and/or installing additional entities such as flexible alternating current transmission systems (FACTS) and energy storage systems (ESSs) such as battery storage systems (BSSs).

Technically, it is true that upgrading the existing electrical networks and/or installing additional entities can help in many aspects, e.g., improving the performance of the power system in terms of stability, reliability, and etc. But, economically, this can lead to incorrect decisions especially for long-term planning of power systems. Of course, in planning and operating power systems many issues, such as renewable energies, BSSs, active and reactive power flows, governmental regulations, etc., need to be considered simultaneously. Otherwise, an incomplete picture of the power system can lead to incorrect decisions. Recently, it is observed that most studies which have been carried out in the field of power transmission and distribution systems used traditional optimal power flow (OPF) methods. Such methods may be suitable for the traditional existing electrical power systems, e.g., the European network as seen in Fig. 1.1, but considering complex bidirectional power flows for future power systems, as seen in Fig. 1.2., new methods are required.

In this work, optimal operation of distribution networks (DNs) with wind and photovoltaic (PV) embedded renewable energy generation and BSSs are considered. A combined active-reactive OPF (A-R-OPF) problem is formulated, i.e., both the active and reactive power distribution will be simultaneously optimized. In particular, the reactive power capability of power conditioning systems (PCSs) of distributed generation (DG) units, BSSs, and on-load-tap-changers (OLTCs) of main transformers (TRs) in DNs is fully employed. It is aimed to minimize energy losses, improve voltage profiles, minimize renewable energy curtailments, and save costs from additional reactive energy imports and/or costs of investments of additional reactive power devices. In addition, different kinds of time horizons (i.e., second, hour, day and year) are considered for carrying out simulation and optimization tasks. Two real case studies have been used to show the effectiveness of the proposed operation strategies. These networks are considered for two different voltage levels, namely low- and medium-voltage (LV and MV) networks.

Chap. 1: Introduction

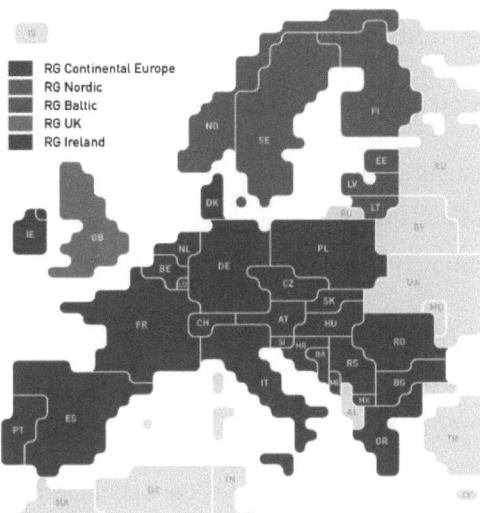

Figure 1.1: European network of transmission system operators for electricity (ENTSO-E) with 5 regional groups in 2013 [127].

Figure 1.2: Vision of European network in 2020 with a high penetration of renewable energies and energy exchanges [127].

1.2 Contributions and Dissertation Structure

The contributions of this dissertation can be summarized as follows:

- Novel mathematical models and a combined problem formulation for A-R-OPF in low- and medium-voltage DNs with embedded renewable energy generation (PV and wind) and BSSs are proposed.

- Development and implementation of dynamic simulators (DSIs) and dynamic optimizers (DOPs) for power flow studies in low- and medium-voltage DNs.

- A novel *flexible* solution strategy for the operation of OLTCs of main TRs in DNs is proposed. The goal is to minimize the total active energy losses during typical season's days.

- Highlighting the impacts of controlling and utilizing DG reactive power capabilities on the operations of DNs. A low-voltage DN with a high penetration of PV systems (PVSs) is considered. In addition, a short-term analysis, i.e., for one year, is made and several novel results are presented.

- A novel solution strategy for the A-R-OPF in medium-voltage DNs with wind and battery storage stations is introduced. In this case, the A-R-OPF is based on a two-tariff price model (for active energy) and one fixed charge/discharge cycle in each day. In addition, two different optimization horizons, namely one-day strategy and multi-day strategy, are considered and compared.

- A novel *flexible* technique for solving the A-R-OPF problem in medium-voltage DNs with wind and battery storage stations. In this case, the A-R-OPF is based on a two-/three-/24-hour-tariff price model (for active energy), a fixed tariff price model (for reactive energy), and one charge/discharge cycle in each day. Here, two different optimization horizons, which are one-day strategy and multi-day strategy, are used to solve the problem.

The structure and chapters of this dissertation is shown in Fig. 1.3.

Chapter 2 gives a literature review of the previous studies related to the topic of this work. Historical timelines are used to help the reader with understanding the events in each phase. It is divided into four phases reviewing the most and relevant publications in each phase. The fourth phase summarizes the contributions of this work.

Chapter 3 presents the modeling procedures made in this work. First, we give a brief background of a modern power system followed by the modeling of: 1) passive DNs (PDNs); 2) active DNs (ADNs); and 3) energy prices.

Chapter 4 analyzes the power flow and OPF in PDNs, i.e., without renewable DG generators and BSSs. The main goal of this chapter is to analyze the operation of PDNs when controlling the voltage amplitude at the secondary bus of the main TRs. In addition, an optimization method is developed to solve this problem as well as results of a case study are presented.

Chapter 5 presents the A-R-OPF method in ADNs with a high penetration of PVSs at the low-voltage level and a high penetration of wind turbines with BSSs at the medium-voltage level. These two voltage levels are separately considered and analyzed. The impacts as well as benefits of

Chap. 1: Introduction

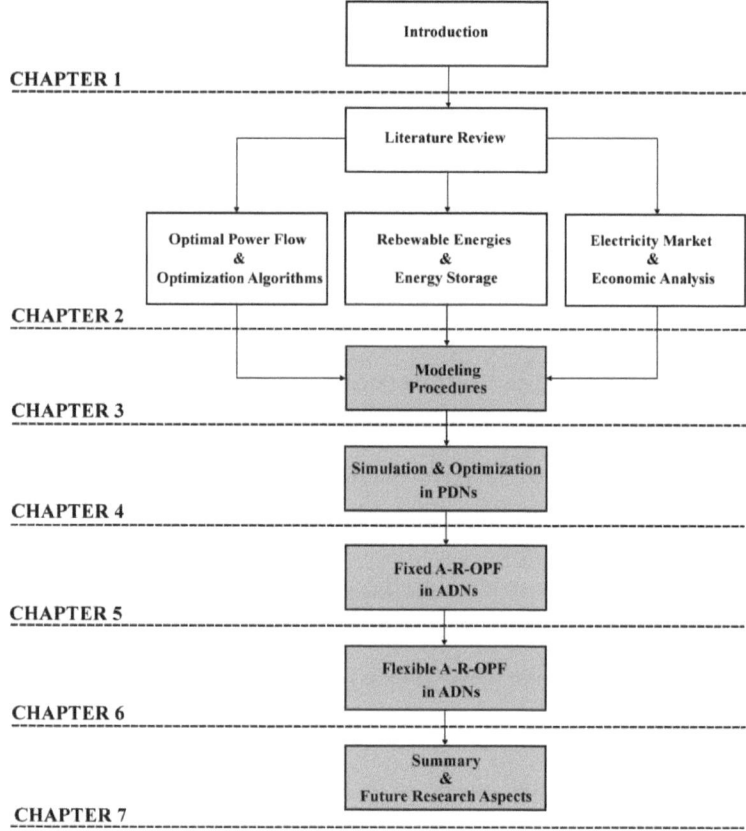

Figure 1.3: Dissertation structure.

considering the reactive power flow in ADNs are shown by applying the A-R-OPF method on two real case studies.

Chapter 6 introduces an extension of the A-R-OPF method. In this chapter, both active and reactive energy prices are considered. Moreover, a flexible battery management system (FBMS) is developed. This is accomplished by optimizing the length (hours) of charge and discharge periods for each day.

Finally, conclusions of this work and future research aspects are given in Chapter 7.

1.3 Software Tools for Simulation and Optimization

The modeling of operations of energy supply networks considered in this work leads to a complex (mixed-integer nonlinear, dynamic, high dimensional and with multiple input disturbances) equation system. For the optimization of such systems, operational constraints (boundaries of all state variables) have to be satisfied.

In this work, model equations are established and different optimizations problems are formulated. The numerical computations are implemented in the MATLAB, Simulink [131], and General Algebraic Modeling System (GAMS) [132]. MATLAB and Simulink environments are well known and widely used for carrying out researches, particularly for calculation and graphical simulation purposes. In contrast, GAMS is a high-level modeling system which has been successfully used to solve models of large-scale systems (e.g., power systems [22]) especially for optimization purposes. Detailed and useful information on the GAMS environment such as the structure of a GAMS model, illustrative examples of model development and some of the GAMS models in power systems literature can be found in [22].

In brief summary, GAMS provides powerful solvers (e.g., IPOPT, BARON, CONOPT, and etc.) for different classes of optimization problems (e.g., linear programming (LP), nonlinear programming (NLP) and mixed-integer nonlinear programming (MINLP), etc.). In this work, CONOPT3 solver is well suited for models with highly nonlinear constraints, and therefore, it is used for solving all formulated optimization problems.

2 Literature Review

A brief historical sequence of some important events and dates related to this work is given in this chapter. The timeline is divided into four phases reviewing some selected and relevant publications. In the first two phases, we shortly give an overview of the development in power systems from year 1800 to 1980, while the third phase gives a more detailed review from year 1980 to 2011. The contributions of this work are summarized in the last phase.

2.1 Direct and Alternating Current

In 1800, Volta [19] announced the invention of the battery which is the backbone of this work. Basically, a battery is a device which converts chemical energy into electric energy and vice versa. In 1831, Faraday [128] discovered the principle of electromagnetism and he was able to build the first electric motor, followed shortly by the first generator and the first transformer. Many inventors tried to improve the basic idea of electromagnetic induction and used magnets to create a flow of current in wires. Pixii, in 1832 [128], was one of those who invented such a machine. In the 1860s and 1870s many inventors sought ways of using Faraday's induction principle to generate electricity mechanically [128]. Accordingly, two kinds of generators emerged. The first type was a generator of direct current (DC) electricity. The second type was a generator of alternating current (AC) electricity.

Edison first invented his electric lamp in 1880 [32] followed by an electric meter (M) in 1881 [33]. He constructed his meters and used them extensively in New York as house meters for the purpose of measuring the "quantity" of electricity passing from the central station to the consumers [55]. After that, he developed a system of underground conductors in 1883 [34] and overhead conductors in 1888 [35] both for DC electrical distribution systems. It is to note that Edison thought of AC as something so "abnormal" as to be impossible of practical use [72].

In 1886, Stanley [100], and in 1887, Westinghouse [120] first invented different electrical TRs for practical usage. In 1888, Tesla [106] invented the induction motor and made some improvements in the transmission of AC power [107]. This basic step in utilization of AC power was a solid foundation of engineering application [73][77]. In 1895, Dobrowolsky [27] invented an apparatus for indicating any lag or advance between the phase of an electric AC current and that of the electromotive force (e.m.f.) pressure of the same current.

Based on AC theory, Browne in 1901 [13] stated that the phase difference of two waves is usually defined as the displacement in degrees between the points where they pass, in the same direction, through zero or their maximum values. If these two waves are assumed sinusoids and one is the e.m.f. and the other is the current, then the cosine of the phase angle is called power factor (PF). It is clear that a PF is unity when the phase angle is zero. But the introduction of Tesla's induction motors in factories [13] has led to a PF considerably *less* than unity. This brought the need to improve PF in AC power systems for many reasons, as shown in the following.

In 1909, Walker [117] made early discussions on the need to improve PF in AC systems. This is because a poor "low" PF makes losses in power systems and limit the capacity of generators. In addition, if the PF is "low", a much larger generator is required than would otherwise be necessary. Moreover, the cables and TRs are much more costly. In 1915, Philip [86] described a method for representing the flow of energy through the main parts of an electric distribution system, i.e., generators, motors, TRs, and transmission lines. Terms such as lagging/leading PF, transmission losses, and power flow directions can also be found in [86]. It was stated that in every AC system two kinds of power (active and reactive power) coexist and flow independently. The two kinds may flow in the same or in opposite directions. Either flow may vary without interfering with the other flow. In 1923, Kapp [63] introduced some apparatus to be used for PF improvement and voltage control, such as static condensers and idle-running synchronous machines. In 1926, an early proposal included the PF as a part of the electricity tariff [25][29]. A three-part tariff was proposed, embracing a periodic charge per kVA of maximum demand, a charge per kWh of energy, and a charge per kVA-hour of lagging wattless component. In 1928, Jansen [58] invented a device to control the voltage of electrical TRs. This device is called OLTC. It is commonly used nowadays by utilities to maintain the voltage in an acceptable range.

In 1930 [52], the first AC network analyzer was installed in the Electrical Engineering Research Laboratory of the Massachusetts Institute of Technology at Cambridge, Mass. In this analyzer, equivalent-π circuits were used to model the lines and transformers of AC networks. In 1933 [60], some operating aspects of reactive power were presented. In addition, a metering technique for keeping track of the flow of reactive power in a complicated power transmission network (TN) which greatly clarifies and simplifies the problem of dispatching reactive power in such a system was also presented. It was stated that dispatching reactive power is necessary to obtain a maximum transmission system capacity as well as for system voltage control.

2.2 Mathematical Formulation of Optimal Power Flow

In 1943 [45], George proposed a first loss model for power networks considering both active and reactive power flows in a TN. It is to note that George used longhand calculations without an aid of computers. In 1946 [108] and 1948 [109], respectively, the concept and the sign of reactive power and its flow were discussed in detail. In 1947 [30], Dunstan was the first who introduced a machine to study and analyze the performance of power networks. It was aimed to set certain principles and methods which facilitate the handling of problems of power flow in networks so that the mathematical solution appears more feasible than the network analyzer. After that, in 1956 [118], Ward and Hale presented a method for solving the power flow problem on digital computers. The power flow problem was solved based on a loop approach.

Briefly, the power flow problem consists of imposing specified power input and voltage amplitude, or active and reactive power input conditions, at the terminals of a passive network. The desired solution will provide complete input and voltage information at the terminal and power flow in each branch of the network. It is noted that the method in [118] was based on an iterative process. In 1957 [48], Glimn and Stagg proposed another iterative method based on the Gauss-Seidel algorithm. It was also aimed to determine the distribution of the system voltage based on a nodal approach. In 1961 [113], Van Ness and Griffin described an elimination method for load flow studies. This method is called later Newton's method [110].

In 1962 [18], Carpentier was the first who formulated a general problem of OPF subject to equality and inequality constraints. In 1963 [98], Smith and Tong presented a method for minimizing power transmission losses by reactive-volt-ampere control. The method was based on varying the voltage of some buses in a transmission power system and then the optimum loss condition was found. In 1967 [110], Tinney and Hart introduced Newton's method to solve the AC power flow. The characteristics of the method such as speed, accuracy, computer requirements and ill-conditioned problems were also described in detail. It was stated that the Newton's method has revealed no ill-conditioned situations.

In 1968 [28], Dommel and Tinney presented a method for solving the power flow problem with control variables such as real and reactive power and transformer ratios automatically adjusted to minimize instantaneous costs or losses. In the same year, a general problem of minimizing the operating cost of a power system by proper selection of the active and reactive productions was formulated by Peschon *et al.* [83]. The formulation was a NLP problem in accordance with a previous work by Carpentier. Peschon *et al.* extended the work in [84] to take sensitivity considerations entering into the optimum dispatching problem, i.e., the sensitivity relations between dependent (state), independent (control) and uncontrollable (parameter) variables. In the same year (1968) [31], Dura proposed the dynamic programming for optimal sizing and allocation of shunt capacitors in radial distribution feeders. In 1969 [92], Sasson presented a unified approach to solve the OPF problem. His approach was essentially based on the Carpentier's formulation. Note that the original method of Carpentier provides exact solutions to AC-OPF problems without any approximation.

Since 1972 [61], power flow optimization methods have been classified in two categories. First, exact methods which take into account both active and reactive power flows in obtaining the solution of OPF problems. Second, approximate methods which achieve simplified representations and possibly computational efficiencies by ignoring either the active or reactive power equations. In 1974 [10], Borkowska presented the third category as a method for solving the power flow problem taking into account uncertainties, such as load nodes, with linear approximations and neglected losses. Exact or approximate stochastic or probabilistic power flow methods have been used for considering uncertainties. In 1979 [38], Felix *et al.* developed a two-stage approach for solving large-scale OPF problems. However, the basic drawback of this approach is that the transmission line flows are not included [38]. In 1982 [97], Shoults and Sun decomposed the OPF problem into two subproblems (P-Problem and Q-Problem), where these two suboptimal problems were solved separately. This method overcomes the method in [38] by considering system constraints. It is noted that Newton's method, used to solve OPF problems, have been faced by so called ill-conditioned situations. Therefore, a study was made in 1982 [112] to deal with load-flow solutions for ill-conditioned power systems by a Newton-like method.

2.3 OPF with Renewable Energies and Storage Systems

Indeed, from 1980 to 2011, a vast number of studies dealing with the OPF problem, renewable energy generation and ESSs for both DNs and TNs have been published. The characteristics of this research period can be summarized as follows:

- Because of integrating a large scale of DG units new problems at both DNs and TNs have been observed and formulated.
- New market models and pricing schemes for both DNs and TNs have been introduced, and therefore, new challenges appear in planning and operating electrical power systems.

- Increasing the need for ESSs, e.g., BSSs, to face the fluctuating of intermittent DG generators, e.g., wind and PV.
- Finally, new methods and techniques have been developed to handle the complexity of planning and operating DNs as well as TNs under the above described circumstances.

A summary of some selected, recent, and relevant literature is given below. A detailed review, however, can be found in surveys in 1991 [57], 1999 part I [75] and part II [76], and the recent state-of-the-art in 2013 [65].

2.3.1 OPF without DG Units

In 2002 [68], Kersting described in detail models and simulation techniques for solving power flow problems in balance and unbalance DNs. In the same year [9], Bakirtzis *et al.* proposed an enhanced genetic algorithm (GA) for the solution of OPF in TNs. In 2004 [1], Acha *et al.* gave in depth the modeling and simulation methods required for a thorough study of the steady-state operation of electrical power systems, especially with FACTS. MATALAB programs were written in [1], in which FACTS controllers such as Static VAR Compensator (SVC), were considered. In 2009 [62], Jong-Young *et al.* proposed a planning method for allocating capacitors in DNs and minimizing energy losses and installation costs. In 2011 [123], Zimmerman *et al.* presented a so called MATPOWER which is an open source MATLAB-based power system simulation package. MATPOWER has the ability to solve large-scale AC- and DC-OPF problems. However, only static operation of power systems have been considered, i.e., with no dynamics in OPF problem formulations. In general, the above reviewed models, simulation and optimization methods did not consider the penetration of DG units.

2.3.2 OPF with DG Units

Nowadays, renewable DG units, as wind and PV are being increasingly considered as attractive, sustainable and green energy sources. This is because of many environmental and economical concerns.

In 1994 [88], Rau and Wan proposed an initial step on optimum allocation of DG units. It was aimed to maximize the benefits of DG units by introducing different objective functions such as network losses and line loads. In 2002 [70], Liew and Strbac proposed different alternative control strategies to increase the penetration of wind-based DG units. Controls such as generation curtailment [111], reactive power absorption and coordinated OLTC were introduced. It was shown that by implementing active network management, the increase of wind-based DG units which can be connected to the existing DNs can be increased considerably. In 2005 [51], Harrison and Wallace proposed a mathematical model for maximizing DG capacity in DNs. One potential criticism of the approach was the use of a single deterministic optimization. In 2008 [115], Viawan addressed in detail the impact of DG units on voltage stability of DNs. It was shown that high penetration of DG units may cause the power to flow reversely from the secondary to the primary side of the transformer. In the same year [82], reactive optimal power flow was considered. It was shown that a considerable reduction of power losses can be achieved.

In 2010 and 2011, Ochoa *et al.* [78][79][80], proposed different models, optimization techniques for assessing and optimizing DG and especially wind-based DG units in DNs. The methods and techniques used can handle the variation of both demand and DG generators. It was aimed to maximize the energy penetration from DG units and meanwhile minimizing energy losses. Minimizing reactive energy import from the TN was also considered. It was shown that a high penetration of renewable-based DG units will lead to high renewable energy curtailments. Economically, such curtailments lead to additional costs and therefore it was important to evaluate the potential of ESSs, such as BSSs to accumulate these curtailments.

2.3.3 Energy Storage for Power Systems

In 1980 [26], Davidson *et al.* gave a review of different types of large-scale electrical ESSs as pumped storage, compressed-air storage, thermal-energy storage, electrochemical battery storage, flywheels and superconducting magnetic energy storage. It was concluded that a storage plant can become increasingly valuable if it is integrated with renewable energy sources. In addition, batteries have the advantage over other storage schemes, especially if they are utilized in small units close to the customers.

Ter-Gazarian [105] stated that BSSs situated close to the consumer are able to smooth the load on the DN, thus decreasing the required capacity of substations. For these reasons, we chose BSSs in this work. BSSs are devices that convert the chemical energy contained in its active materials directly into energy by means of an electrochemical oxidation-reduction (redox) reaction [89]. In 1990 [116], Walker described a bidirectional 18-pulse voltage source converter utilizing gate turn-off thyristors (GTOs). The converter was rated with 10 MVA to connect a BSS to a utility grid. It was demonstrated that GTO thyristor power circuit is capable to provide independent fast-response control of active and reactive power. In 1996 [74], Miller *et al.* described the design and commissioning of a 5 MVA, 2.5 MWh BSS. This BSS, e.g., was placed in service at the GNB Battery Recycling Plant, Vernon, California for two main functions. First, a BSS permits critical loads up to 2.5 MW to operate for up one hour when external disturbances occur. Second, it was used to supply energy to the plant during peak load periods. In 2001 [90], Ribeiro *et al.* gave a detailed review of BSSs's capabilities, such as dynamic and transient stability, voltage support, area control and frequency regulation, transmission capability and power quality improvement.

In 2003 [93], many energy storage technologies were examined for three application categories (bulk energy storage, DG, and power quality) with significant variations in discharge time and storage capacity. In 2005 [20][21], Chacra *et al.* studied the impact of energy storage costs on economical performance of a distribution substation, where the benefit and cost of installing an ESS were evaluated. Two types of ESS technology were chosen in the evaluation process, namely, Vanadium Redox Batteries (VRBs) and Polysulfied-Bromine (PSB). It was shown that PSB batteries are likely to be more cost-effective than VRB. Many energy storage technologies were examined in [93] for three application categories (bulk energy storage, DG and power quality) with significant variations in discharge time and storage capacity. In 2008 [87], Poonpun and Jewell made a cost analysis considering life-cycle of grid-connected electric energy storage. They showed that the length of discharge cycle (not the depth of discharge (DoD)) has an impact on the cost added to the electricity cost.

Recently [114], a study was made on ESSs in power supply systems with a high share of renewable energy sources. In 2009 [103] and 2010 [104], Teleke *et al.* focused on developing an optimal control method which integrates a BSS with a large wind farm. It was aimed to have the BSS to provide as much smoothing as possible, so that the wind farm can be dispatched on an hourly basis based on forecasted wind conditions.

More details on BSSs's technologies and applications can be found in surveys, e.g., in 2010 by Ahlert [3].

It is now obvious that from all types of ESSs, BSSs are expected to have a significant role in the future energy networks, and therefore, we will focus on it in this work.

2.3.4 OPF with DG Units and BSSs

In 2010 [7], Atwa and El-Saadany proposed a technique for sizing and optimal allocating BSSs in DNs with a high penetration of wind-based DG turbines. Different types of BSSs were evaluated and an economical analysis has been done. The BSSs's reactive power potential was not considered. In 2010 [46], Geth *et al.* proposed an optimization method based on a trade-off between different objectives for siting and sizing BSSs in DNs. It was shown that installing more BSSs would lower the probability of voltage violation. The authors in [46] proposed a techno-economical model for BSSs in [47][102]. The work in [47] was partially based on [3] where a depreciation cost of cycling a BSS was used. However, in [3], [46] and [47] the opportunity cost of reactive power during cycling a BSS was not included.

In 2011 [81], Oh proposed modeling storage devices in the OPF framework to take the TN into consideration. In [81] the problem was formulated using a lossless DC-OPF model, and therefore, losses, voltage, and reactive power are not considered. The OPF problems in [7] and [81], were formulated for medium-voltage DNs and high-voltage (HV) TNs, respectively. In 2011 [23], Chen *et al.* proposed a method for sizing and economic analysis of ESSs in low-voltage DNs. A variety of DG sources have been considered as sources with unity PF.

2.3.5 Traditional Electricity Market

In 1982 [17] Caramanis *et al.* and in 1988 [94], Schweppe *et al.* presented a complete framework for so called "spot pricing of electricity". It was stated in [94] that "Electric energy must be treated as a commodity which can be bought, sold, and treated, taking into account its time- and space-varying values and costs". In addition, an Energy Marketplace was defined by "The buying and selling of electric energy between independent customers and a regulated or government owned utility." In brief summary, the hourly spot price for electric energy was the main foundation in [94].

Later, locational marginal price (LMP) was applicable by independent system operators (ISOs) of TNs. Basically, LMP ($/MW) is composed of three components: System Marginal Energy Cost (SMEC), Marginal Cost of Congestion (MCC), and Marginal Cost of Losses (MCL). In each hour of the day-ahead market of energy a system operator (e.g., California Independent System Operator (CISO) [129], Independent System Operator New England (ISO-NE) [130], and etc.), calculates the LMP for each node (bus) in a specific transmission zone. Usually, the first component is called market clearing price (MCP), which can be determined based on *bids* (to buy energy) and *offers* (to sell energy) into the market from dispatchable facilities. Note that LMP = MCP = SMEC in power systems with no congestions and ignoring losses, but LMP can be different for different buses when there are congestion and losses. Briefly, the SMEC component reflects the marginal cost of providing energy from a designated reference location [129].

In 2002 [95], Shahidehpour *et al.* provided detailed representations on transmission congestion management and pricing. It was stated that MCP may be different for various transmission zones, but it is the same within a zone. For example, if we have two zones and the ISO detects a congestion along the transmission path connecting these zones, then the schedules in each zone will be adjusted to relieve the congestion. This can lead to two different MCPs which can be denoted by zonal MCP (ZMCP), i.e., ZMCP-1 for zone 1 and ZMCP-2 for zone 2. In practice, interconnection lines (see Fig. 2.1) allow electricity to be imported into and exported out of a certain zone [95][130].

2.3.6 Electricity Market with DG Units and BSSs

In the last two decades, DG units were expensive in comparison with other conventional energy sources, such as nuclear power, gas and coal. Therefore, it was necessary to support them by some governmental regulations to effectively accelerate its integration.

In Germany, for example, the feed-in-tariff (FIT) was first brought in as part of the Stromeinspeisungsgesetz (StrEG), and since 2000 has been part of the Erneuerbare-Energien-Gesetz (EEG) [71]. In Ontario, Canada, the FIT was introduced as part of the Green Energy and Green Economy Act (GEGEA) of 2009, which replaced an earlier standards offer program for renewable electricity [71]. It is worth mentioning to note, however, that the above reviewed acts and regulations are being dynamically changed and updated. Therefore, researchers in the field of planning and operations of power systems need to keep all these updates in mind when taking their decisions.

Chap. 2: Literature Review

Figure 2.1: Illustration of a control center in electric power industry [130].

2.4 Flexible A-R-OPF with Renewable Energies and BSSs

It is clearly seen from the literature above that electrical power systems become more and more complex. Therefore, it is important to develop new models, methods as well as tools to deal with these complexities. During the last four years, new contributions have been made from this work based on the above significant works. A short description of these contributions is given here, whereas more details are given in the coming chapters.

In 2011 [39], the reactive power capability of a wind-battery station in electricity market was investigated. The objective was to balance the maximization of the total revenue and meanwhile the maximization of the amount of available reactive power. It was shown that a large amount of reactive power can be achieved by an optimal operation strategy. In 2012 [40], a combined problem formulation for A-R-OPF in DNs with embedded wind generation and BSSs was proposed. It was shown that using A-R-OPF a huge reduction in both energy losses and reactive energy can be achieved. The problem was formulated as a large-scale nonlinear optimization problem in which the lengths of charge and discharge periods for BSSs were fixed in daily operations. In 2012 [42], the method in [40] was extended to a flexible operation strategy for BSSs. A complex MINLP problem was formulated and solved. In [40] and [42], DG units were considered to work at a constant unity PF. In 2012 [41], a new mathematical model for utilizing both active and reactive power capability of DG units was introduced. In 2012 [43] and 2013 [44], initial studies have been made for operating TNs and planning DNs with considering the impacts of bidirectional power flow between TNs and DNs.

3 Modeling Procedures

3.1 Background

A modern power system can be divided into four main parts: 1) generating plants (Gs); 2) transmission system (TS); 3) distribution system (DS); and 4) final consumers. As shown in Fig. 3.1, the TS and DS consist of electrical networks which work with multi-voltage levels, namely, very high-voltage (VHV) (e.g., 220 and 380 kV), high-voltage (HV) (e.g., 66 and 110 kV), medium-voltage (MV) (e.g., 10 and 20 kV) and low-voltage (LV) (e.g., 0.4 kV) [54]. It is convenient to call TS networks by TNs and DS networks by DNs. Typically, these networks are connected together by TRs for voltage adaptation. Here, we call the primary side of a TR by S_0 (input bus) and to the secondary side by S_1 (output bus). Usually, TRs are equipped by OLTCs to regulate and hold the voltage at a regulated bus in an acceptable range [15]. In this work, the bus S_1 (called slack/infinite bus) is selected as a regulated bus.

Typically, generation companies produce electricity, while operators/companies of the TS/DS transport/distribute it to final consumers. This paradigm leads to unidirectional power flows, in which the downstream DNs import active and reactive energy from the upstream TNs, as illustrated in Fig. 3.1. In this way, conventional DNs are usually called PDNs [24]. In contrast, downstream ADNs (with DG units and/or BSSs) have the possibility to export active and reactive energy to upstream TNs [40][42], leading to bidirectional power flows, as seen in Fig. 3.2.

Chap. 3: Modeling Procedures

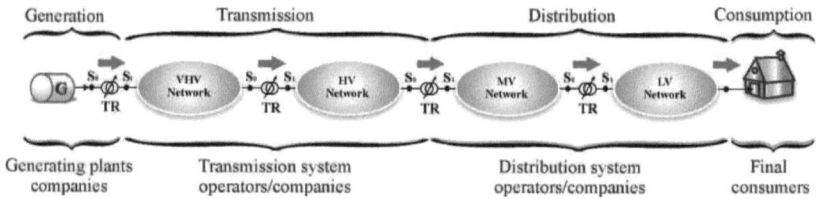

Figure 3.1: Simplified structure of the modern power system.

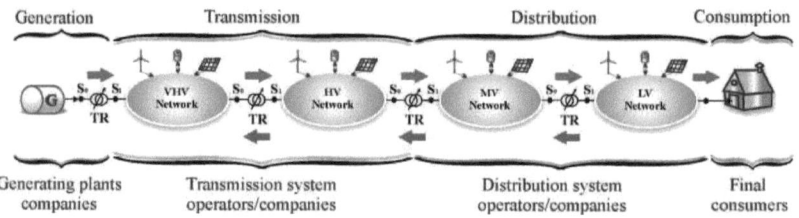

Figure 3.2: Simplified structure of the future power system.

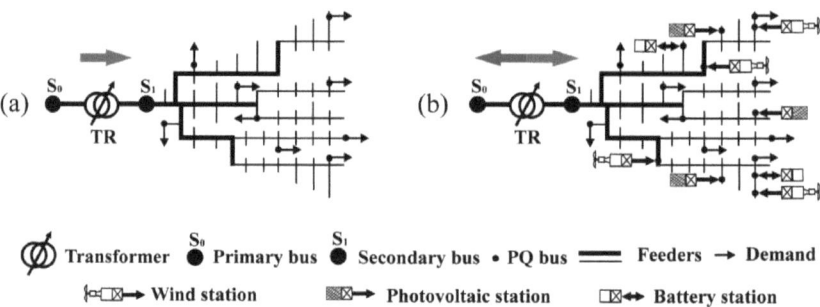

Figure 3.3: (a) A typical PDN [121] (unidirectional power flow). (b) An ADN [42] (bidirectional power flow).

A typical PDN layout is depicted in Fig. 3.3(a) where the line thickness indicates feeder capacity. Note that an operating point of PDNs lies in quadrant 1 (green area), as seen in Fig. 3.4.

Chap. 3: Modeling Procedures

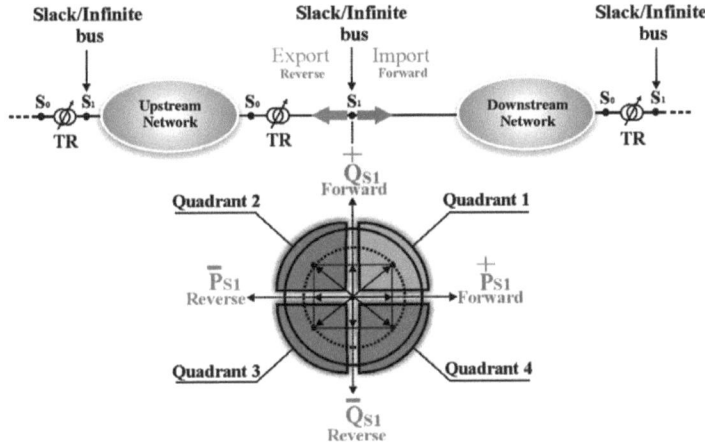

Figure 3.4: Active-reactive power flow direction definitions for upstream and downstream networks [101].

However, an operating point of an ADN can lie in any of the second, third, and fourth quadrant, as shown in Fig. 3.4 (red area).

Since it is critical to consider the whole power system, seen in Fig. 3.2, at the same time, the focus in this work will be on the DS. Therefore, a typical low-voltage DN and typical medium-voltage DN are separately modeled in this work. In the next sections, we consider the modelling of: 1) PDNs; 2) ADNs; and 3) energy prices, respectively.

3.2 Modeling of Passive Distribution Networks

As described in Fig. 3.3(a), a typical PDN comprises of a variety of demands at different buses, feeders, and a TR with an OLTC. The mathematical models used in this work to describe all these entities are given in the following. In addition, the nonlinear power equations are described and the Newton-Raphson method for solving the power flow equations is presented.

Chap. 3: Modeling Procedures

3.2.1 Load Model and Bus Types

The active P_d and reactive Q_d load or demand profiles, as shown in Fig. 3.5, are assumed in this work to follow the IEEE-RTS typical season's days as given in [50][7]. These profiles are calculated based on the hourly load data and given as a percentage of the annual peak load (see Appendix A, Table A.1). The secondary bus of the main TR in PDNs and ADNs is selected as slack bus, whereas other buses are considered as PQ buses. This is valid for low- and medium-voltage networks.

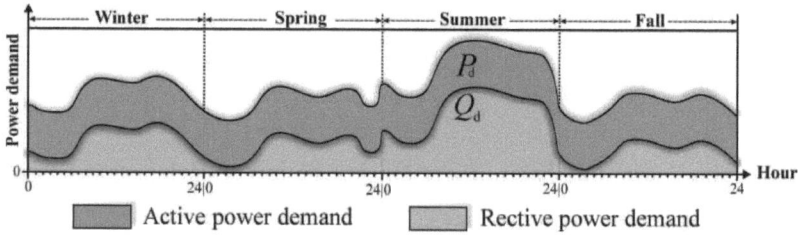

Figure 3.5: Typical daily active (P_d) and reactive (Q_d) power demand.

3.2.2 Feeder Model of Distribution Networks

The modeling of distribution overhead and underground line segments is a critical step in the analysis of a distribution feeder [68]. Typically, a distribution feeder can be represented by an equivalent-π circuit with a series impedance ($R_1 + j\, X_1$) and a shunt admittance ($G_1 + j\, B_1$), as shown in Fig. 3.6. The series impedance includes the total resistance R_1 and inductive reactance X_1 of the line segment, while the shunt admittance (divided between its shunt arms) includes the total conductance G_1 and capacitive susceptance to natural B_1. The total shunt admittance or a part of it can be neglected because its impact is very small comparing with the series impedance [36].

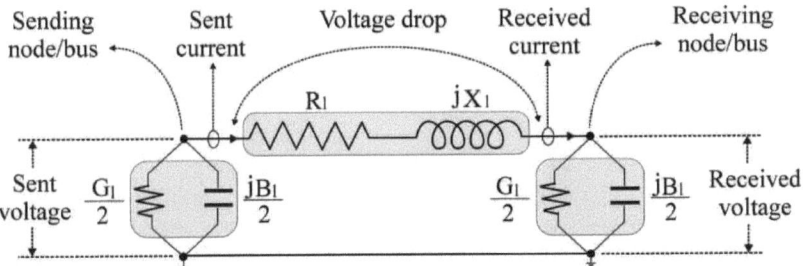

Figure 3.6: Illustration of a single phase equivalent-π circuit of a feeder.

In this work, the total shunt admittance is neglected in the low-voltage network and only the total conductance is neglected in the medium-voltage network.

As seen from Fig. 3.6, the sent current is equal to the received current. This current causes a voltage drop between the sending and receiving sides. The voltage drop depends on many factors, e.g., the amplitude and the direction of the current passing through a line segment.

3.2.3 On-Load Tap Changer Transformer

In 1984 [15], Calovic considered the problem of modeling and analysis of OLTC transformer control systems. A nonlinear model was derived for analysis of voltage and reactive power control applications considering mid-term and long-term dynamics and steady-state behavior of power systems. The equivalent circuit of an OLTC TR is given as a cascade of two *ideal* TRs, as depicted in Fig. 3.7 [15]. Here, a:1 is the nominal tap-ratio of the autotransformer, $N_1:N_2$ is the nominal turn-ratio of the OLTC transformer, Z_t: is the equivalent transformer impedance consisting of R_t: the transformer nominal resistance and X_t: the transformer nominal leakage reactance. This OLTC transformer model is used in this work for analysis purposes.

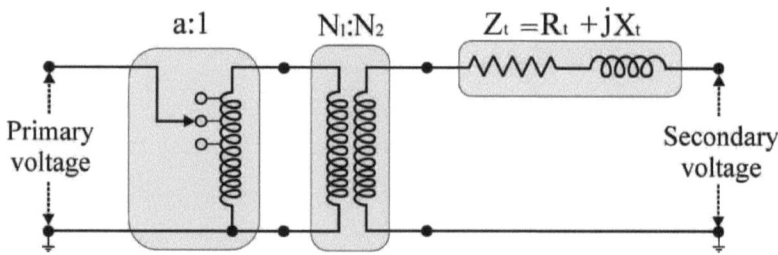

Figure 3.7: Simplified block diagram of a single phase of an OLTC transformer.

3.2.4 Power Flow in PDNs

In [36][1][68], an electrical power network consists of a number of nodes (buses) N connected together with a number of lines N_{line}, as seen in Fig. 3.8. Based on the theory of power flow, an instantaneous power balance must be satisfied at each bus i, i.e., the power produced from a conventional generator (CGE) (i.e., active power $P_g(i,h)$ and reactive power $Q_g(i,h)$) minus the power consumed by a demand (DEM) (i.e., active power $P_d(i,h)$ and reactive power $Q_d(i,h)$), connected to a bus i at hour h should be equal to the power exchange with other connected buses (i.e., active power $P(i,h)$ and active power $Q(i,h)$). The nonlinear power flow equations can (in general) be written by $\mathbf{F}(\mathbf{X}) = 0$, as given in [1]. The Newton-Raphson method is the de facto standard for solving the nonlinear power flow equations [1]. This method is used in this work to solve the power flow equations for simulation purposes.

3.2.5 Newton-Raphson Method

Basically, the Newton-Raphson method has been successfully used to solve a set of non-linear equations for large-scale power flow studies. A dynamic simulator, based on the Newton-Raphson method [1], is developed in this work.

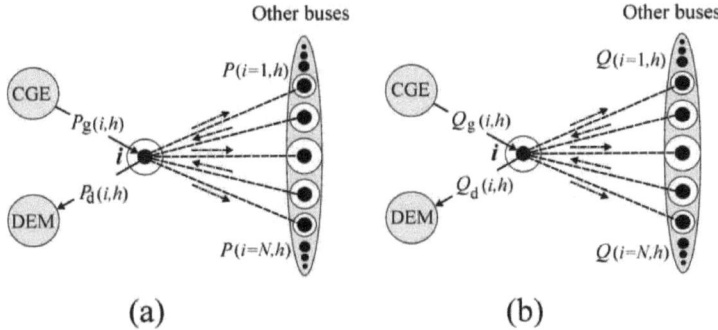

Figure 3.8: Power balance at bus i during hour h: (a) active power (b) reactive power.

This is necessary to dynamically evaluate power systems and analyze its characteristics, such as the power flow Jacobian matrix for well- and ill-conditioned power systems [112][40].

The Newton-Raphson algorithm in is based on an iterative process to solve a set of non-linear algebraic equations:

$$\left.\begin{array}{l} f_1\left(x_1^{(h)}, x_2^{(h)}, \ldots, x_{NE}^{(h)}\right) = 0, \\ f_2\left(x_1^{(h)}, x_2^{(h)}, \ldots, x_{NE}^{(h)}\right) = 0, \\ \quad \cdot \\ \quad \cdot \\ \quad \cdot \\ f_{NE}\left(x_1^{(h)}, x_2^{(h)}, \ldots, x_{NE}^{(h)}\right) = 0, \end{array}\right\}, \text{ or } \mathbf{F}\left(\mathbf{X}^{(h)}\right) = 0 \quad (3.1)$$

where $\mathbf{F}$ represents the set of NE nonlinear equations, $\mathbf{X}^{(h)}$ is the vector on NE unknown state variables at each time interval h. To solve (3.1), the first order of a general Taylor series expansion of $\mathbf{F}(\mathbf{X}^{(h)})$ around an initial guess value $\mathbf{X}^{((it-1),h)}$ is used. Here, the value of the $\mathbf{F}(\mathbf{X}^{(it,h)})$ at iteration $it \in (1, 2,\ldots, it_{max})$ and time interval $h \in (1, 2,\ldots, T_{final})$ can be written as follows

$$\mathbf{F}\left(\mathbf{X}^{(it,h)}\right) \approx \mathbf{F}\left(\mathbf{X}^{((it-1),h)}\right) + \mathbf{J}\left(\mathbf{X}^{((it-1),h)}\right)\left(\mathbf{X}^{(it,h)} - \mathbf{X}^{((it-1),h)}\right), \quad (3.2)$$

where $\mathbf{J}(\mathbf{X}^{((it-1),h)})$ is the Jacobian matrix of $\mathbf{F}(\mathbf{X}^{(h)})$ evaluated at $\mathbf{X}^{((it-1),h)}$. If it is assumed that $\mathbf{X}^{(it,h)}$ is sufficiently close to the solution $\mathbf{X}^{(*,h)}$ then $\mathbf{F}(\mathbf{X}^{(it,h)}) \approx \mathbf{F}(\mathbf{X}^{(*,h)}) = 0$. Thus, Eq. (3.2) can be expressed by

$$\underbrace{\begin{bmatrix} f_1(\mathbf{X}) \\ f_2(\mathbf{X}) \\ \vdots \\ f_{NE}(\mathbf{X}) \end{bmatrix}}_{\mathbf{F}\left(\mathbf{X}^{(it,h)}\right)} = \underbrace{\begin{bmatrix} f_1(\mathbf{X}) \\ f_2(\mathbf{X}) \\ \vdots \\ f_{NE}(\mathbf{X}) \end{bmatrix}}_{\mathbf{F}\left(\mathbf{X}^{((it-1),h)}\right)} + \underbrace{\begin{bmatrix} \frac{\partial f_1(\mathbf{X})}{\partial x_1} & \frac{\partial f_1(\mathbf{X})}{\partial x_2} & \cdots & \frac{\partial f_1(\mathbf{X})}{\partial x_{NE}} \\ \frac{\partial f_2(\mathbf{X})}{\partial x_1} & \frac{\partial f_2(\mathbf{X})}{\partial x_2} & \cdots & \frac{\partial f_2(\mathbf{X})}{\partial x_{NE}} \\ \vdots & \vdots & & \vdots \\ \frac{\partial f_{NE}(\mathbf{X})}{\partial x_1} & \frac{\partial f_{NE}(\mathbf{X})}{\partial x_2} & \cdots & \frac{\partial f_{NE}(\mathbf{X})}{\partial x_{NE}} \end{bmatrix}}_{\mathbf{J}\left(\mathbf{X}^{((it-1),h)}\right)}_{\mathbf{X}=\mathbf{X}^{((it-1),h)}} \underbrace{\begin{bmatrix} x_1^{(it,h)} - x_1^{((it-1),h)} \\ x_2^{(it,h)} - x_2^{((it-1),h)} \\ \vdots \\ x_{NE}^{(it,h)} - x_{NE}^{((it-1),h)} \end{bmatrix}}_{\mathbf{X}^{(it,h)} - \mathbf{X}^{((it-1),h)}} = 0.$$

(3.3)

Let $\Delta\mathbf{X}^{(it,h)} = \mathbf{X}^{(it,h)} - \mathbf{X}^{((it-1),h)}$ be the correction vector or mismatches, then from Eq. (3.3) we yield

$$\Delta\mathbf{X}^{(it,h)} = -\mathbf{J}^{-1}\left(\mathbf{X}^{((it-1),h)}\right)\mathbf{F}\left(\mathbf{X}^{((it-1),h)}\right). \tag{3.4}$$

After solving (3.4), the initial guesses for the next iteration will be updated by the following equation

$$\mathbf{X}^{(it,h)} = \mathbf{X}^{((it-1),h)} + \Delta\mathbf{X}^{(it,h)}. \tag{3.5}$$

This process will be repeated many times as required until a norm of the correction vector $\Delta\mathbf{X}^{(it,h)}$ is less than or equal to a predefined small tolerance, e.g., $\varepsilon = 10^{-9}$. Typically, four to five iterations [1] are needed by the Newton-Raphson method to solve the power flow problem. But if some convergence problems take place the algorithm is terminated by a maximum number of iterations, e.g., $it_{max} = 100$. The same process will be also repeated for each time interval h till a maximum number of time intervals T_{final}. Here, the selection of T_{final} is flexible, e.g., T_{final} = 96 hour, 5760 minute or 345600 second, as shown later.

If the above general mathematical derivatives are applied to the power flow problem, then the correction vector in Eq. (3.4) is called the power mismatch equations as follows [1]

$$\underbrace{\begin{bmatrix} \Delta P \\ \Delta Q \end{bmatrix}^{(it,h)}}_{F\left(X^{((it-1),h)}\right)} = - \underbrace{\begin{bmatrix} \frac{\partial P}{\partial \theta} & \frac{\partial P}{\partial V}V \\ \frac{\partial Q}{\partial \theta} & \frac{\partial Q}{\partial V}V \end{bmatrix}^{(it,h)}}_{J\left(X^{((it-1),h)}\right)} \underbrace{\begin{bmatrix} \Delta \theta \\ \frac{\Delta V}{V} \end{bmatrix}^{(it,h)}}_{\Delta X^{(it,h)}}, \qquad (3.6)$$

where ΔP and ΔQ are the active and reactive power mismatches, respectively, X represents the unknown nodal voltage amplitudes V and phase angles θ. The detailed and entire mathematical formulations of the Jacobian matrix can be found in [1].

3.3 Modeling of Active Distribution Networks

This section describes the mathematical models used for REGs (wind and PV) and BSSs. In addition, new mathematical power flow equations with REGs and BSSs are introduced. As described above, an ADN can import and export active and reactive power through the slack bus. However, predefined bounds on bidirectional power flows are necessary [7][40], as explained later.

3.3.1 Wind Power

Wind power is related to wind speed and can be evaluated by [53]

Chap. 3: Modeling Procedures

$$P_w(v) = \begin{cases} 0, & 0 \leq v \leq v_{ci} \\ P_W((v-v_{ci})/(v_r - v_{ci})), & v_{ci} \leq v \leq v_r \\ P_W, & v_r \leq v \leq v_{co} \\ 0, & v_{co} \leq v \end{cases} \quad (3.7)$$

where the power curve parameters of the wind turbine are usually $v_{ci} = 4$, $v_r = 14$, $v_{co} = 24$ [7]. The above Eq. (3.7) represents a linear relationship between the wind speed and output power of a wind station.

In this work, hourly wind speed data for a year from a city in Germany are used as wind power penetration. Generally, in mid- and long-horizon studies a wind park connected to a network at bus i can be dispatched on an hourly basis [103][104][39], i.e., hourly constant wind active power output $P_w(i,h)$. Here, h denotes a time interval in hour. Because wind-based REGs produce energy depending on wind speeds, it is necessary to introduce a curtailment factor $\beta_{c.w}(i,h)$ at each wind park connected at a specific bus i during hour h. This factor is multiplied by the hourly constant wind active power output $P_w(i,h)$ to reduce its output and ensure a feasible solution [40]. In other words, $\beta_{c.w}(i,h) = 1$, if no wind power will be curtailed, otherwise $\beta_{c.w}(i,h) < 1$.

From another perspective, the PF of a wind park is controllable from 0.95 (inductive) to 0.95 (capacitive) [39]. In this work, the PF of wind parks is assumed constant (PF=1), i.e., a wind park neither absorbs nor provides reactive power. This assumption is made for comparison purposes.

3.3.2 Photovoltaic Power

In general, a PVS consists of two main parts: 1) a PV panel to collect energy from sunlight to produce a DC and 2) a Photovoltaic-PCS (PV-PCS) to convert this current to a suitable AC.

Chap. 3: Modeling Procedures

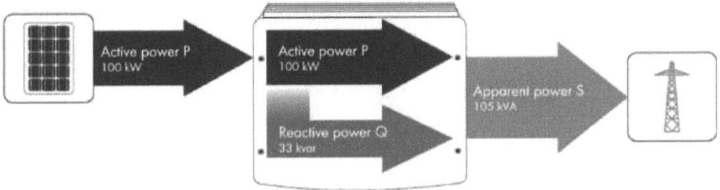

Figure 3.9: Influence of the PF=0.95, or reactive power share on the active power of inverter [126].

A PV-PCS can produce both active and reactive power which can be controlled independently [11][14][69]. In practice [126], the AC power of most inverters is generally given as apparent power in VA. At PF = 1, active and apparent power are equal. Hence, there is no need to make any provision for oversizing in the design of the inverter. However, if the power factor is specified, provisions are typically made for oversizing inverters. This is made to allow producing active and reactive power, as shown in Fig. 3.9. In this work, real data of a PVS from a city in Germany is used as PV power penetrations to define a PV power profile.

The power capability of a PV-PCS is depicted in Fig. 3.10., while the conceptual relation between the active and reactive power of a PV-PCS is depicted in Fig. 3.11. The relations between active and reactive power can be described as follows

$$S_{PCS.pv}(i,h) = \sqrt{\left(P_{pv}(i,h)\beta_{c.pv}(i,h)\right)^2 + \left(Q_{disp.pv}(i,h)\right)^2}, \quad (3.8)$$

$$Q_{disp.ava.pv}(i,h) = \pm\sqrt{\left(S_{PCS.max.pv}(i)\right)^2 - \left(P_{pv}(i,h)\beta_{c.pv}(i,h)\right)^2}, \quad (3.9)$$

$$\left(P_{pv}(i,h)\beta_{c.pv}(i,h)\right)^2 + \left(Q_{disp.pv}(i,h)\right)^2 \leq \left(S_{PCS.max.pv}(i)\right)^2, \quad (3.10)$$

$$-S_{PCS.max.pv}(i) \leq Q_{disp.pv}(i,h), \quad (3.11)$$

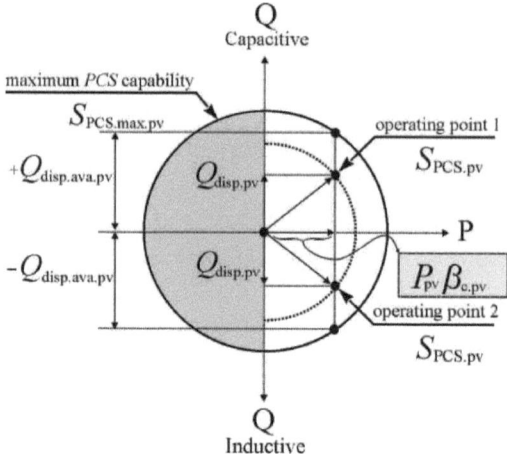

Figure 3.10: Active and reactive power capability of an ideal PV-PCS.

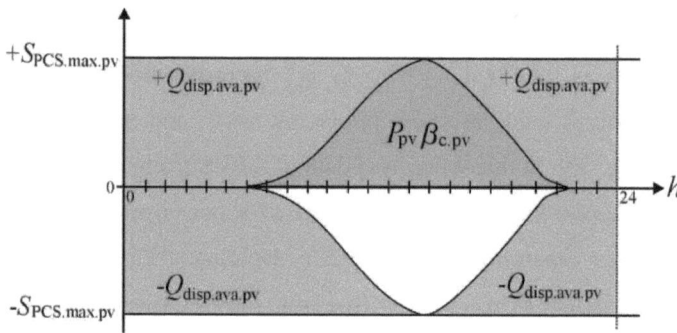

Figure 3.11: Conceptual relation between active and reactive power of an ideal PV-PCS.

where $S_{PCS.pv}$ (i,h) is the apparent power of a PV-PCS at bus i during hour h, P_{pv} (i,h) is a defined PV power profile generated from a PVS at bus i during hour h, $Q_{disp.pv}(i,h)$ is the reactive power of a PV-PCS at bus i during hour h, $S_{PCS.max.pv}$ (i) is the maximum apparent power capability of a PV-PCS at bus i, $Q_{disp.ava.pv}$ (i,h) is the available reactive power of a PV-PCS at bus i during hour h, respectively.

Note that the capability of a PV-PCS is explored by introducing a curtailment factor $\beta_{c.pv}(i,h)$ at each PVS connected at a specific bus i during hour h. In this way, a feasible solution can be ensured, i.e., to spill a part of PV energy out when system constraints will be violated. In other words, $\beta_{c.pv}(i,h) = 1$, if no PV power will be curtailed, otherwise $\beta_{c.pv}(i,h) < 1$.

3.3.3 Battery Storage

Several forms of energy can be defined in physics, and in transforming energy from one form to another, there is sometimes a mechanism for storage [105]. Chemical energy is one of those forms. As discussed in [40], there exists a high potential to implement BSSs in DNs, when: 1) a high penetration of wind power is available; 2) the price model is a multi-tariff system; and 3) an optimal compromise between the active and reactive power is obtained. In addition, if BSSs are considered in a power system, then the potential of BSSs should be fully exploited to compensate its installation costs. This is achieved in this work by developing new operation strategies for BSSs. Generally, a BSS consists mainly of two main parts: 1) a Battery-PCS (B-PCS) unit and 2) a storage unit [39].

A *B-PCS unit* is a voltage source inverter designed to operate as either an inverter when discharging or as a rectifier when charging the battery [74]. It was shown in [74]and [116] that a B-PCS permits a BSS to generate both active and reactive power in all four quadrants as indicated by the capability curve in Fig. 3.12, where two different operating points (i.e., from generating point of view) are depicted.

At operating point 1 the BSS is being discharged with a lagging power factor (supply reactive power), while at operating point 2 the BSS is being charged with a leading power factor (absorbing reactive power).

Chap. 3: Modeling Procedures

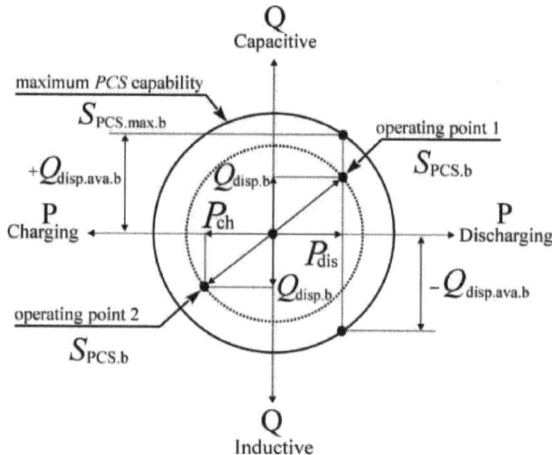

Figure 3.12: Active and reactive power capability of a B-PCS.

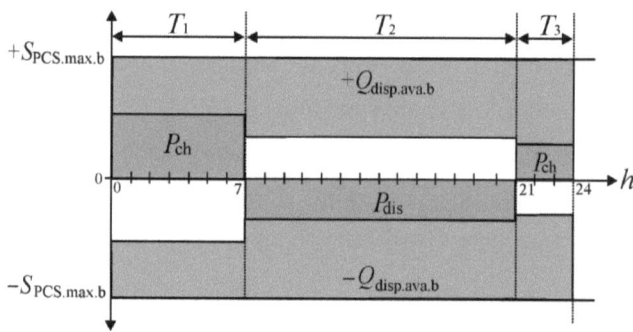

Figure 3.13: Conceptual relation between active and reactive power of a B-PCS.

A *B-PCS unit* is also capable of independent and rapid control of active and reactive power [116]. Based on this capability, three independent control variables for each B-PCS are defined in this work. Two control variables are defined for active power charging and discharging, respectively, and one control variable for reactive power dispatching. It should be noted however that a BSS can only be either charged or discharged for active power at a time point.

On the other hand, energy prices are low during night hours and high during day hours. This is in accordance with typical demand profiles depicted in Fig. 3.5, i.e., the demand is low during night hours and high during day hours. Based on these facts, the control variable for active power charging are taken to be zero during on-peak periods, i.e., during time period T_2 (from hour 7-to-21) as shown in Fig. 3.13. In contrast, the control variable for discharging are taken zero during the off-peak periods, i.e., during time periods T_1 and T_3 (from hour 0-to-7 and from hour 21-to-24). This can be described by

$$\begin{cases} P_{ch}(i,h) = 0, & h \in T_2 \\ P_{dis}(i,h) = 0, & h \in T_1, T_3 \end{cases} \quad (3.12)$$

where $P_{ch}(i,h)$ and $P_{dis}(i,h)$ are the active power charge and discharge of a BSS i during hour h, respectively. The apparent power $S_{PCS.b}(i,h)$ and the available reactive power $Q_{disp.ava.b}(i,h)$ of a B-PCS i during hour h, respectively, can be represented by

$$S_{PCS.b}(i,h) = \begin{cases} \sqrt{\left(P_{ch}(i,h)\right)^2 + \left(Q_{disp.b}(i,h)\right)^2}, & h \in T_1, T_3 \\ \sqrt{\left(P_{dis}(i,h)\right)^2 + \left(Q_{disp.b}(i,h)\right)^2}, & h \in T_2 \end{cases} \quad (3.13)$$

$$Q_{disp.ava.b}(i,h) = \begin{cases} \pm\sqrt{\left(S_{PCS.max.b}(i)\right)^2 - \left(P_{ch}(i,h)\right)^2}, & h \in T_1, T_3 \\ \pm\sqrt{\left(S_{PCS.max.b}(i)\right)^2 - \left(P_{dis}(i,h)\right)^2}, & h \in T_2 \end{cases} \quad (3.14)$$

where $Q_{disp.b}(i,h)$ is the reactive power dispatch of a BSS i during hour h, $S_{PCS.max.b}(i)$ is the upper bound of apparent power of a BSS i. These relations can be clearly explained with Fig. 3.12 and Fig. 3.13. According to the capability limitation of the B-PCS, as shown in Fig. 3.13, following inequalities must be held

$$\left(P_{dis}(i,h)\right)^2 + \left(Q_{disp.b}(i,h)\right)^2 \leq \left(S_{PCS.max.b}(i)\right)^2, \quad (3.15)$$

$$\left(P_{ch}(i,h)\right)^2 + \left(Q_{disp.b}(i,h)\right)^2 \leq \left(S_{PCS.max.b}(i)\right)^2, \quad (3.16)$$

$$P_{ch}(i,h) \geq 0, \quad (3.17)$$

$$P_{dis}(i,h) \geq 0, \quad (3.18)$$

$$-S_{PCS.max.b}(i) \leq Q_{disp.b}(i,h). \quad (3.19)$$

A battery *storage unit* is used to store the energy in the form of chemical energy [39]. The hourly energy balance in each storage unit can be written as

$$E(i,h+1) - E(i,h) - \eta_{ch} P_{ch}(i,h) + P_{dis}(i,h)/\eta_{dis} = 0, \quad (3.20)$$

where $E(i,h)$ is the energy level in a BSS i during hour h. In this work, we assume a fixed value for both charge η_{ch} and discharge η_{dis} efficiencies with a value of 0.77 [7]. Since we consider a time horizon (e.g., T_{final} =24h) for optimization, it is commonly recognized that the energy level in the storage unit at the final time point should be equal to that at the initial time point. Thus we have the following equation for a time horizon

$$E(i, T_{initial}) = E(i, T_{final}), \quad (3.21)$$

where $T_{initial}$ is the initial time in a time horizon. Moreover, upper and lower bound of the storage units should be satisfied, i.e.,

$$E_{min}(i) \leq E(i,h) \leq E_{max}(i). \quad (3.22)$$

Here we assume the lower and upper bound are 20% and 90% of the installed capacity of the storage units, respectively.

3.3.4 Power Flow in ADNs with Battery Storage

In this work, new power flow equations considering REGs and BSSs, as seen in Fig 3.14, are introduced. The proposed power/energy balance (see Fig. 3.14) leads to complex power and energy equations as follows

$$V_e(i,h)\sum_{\substack{j=1\\j\neq i}}^{N}\left(G(i,j)V_e(j,h)-B(i,j)V_f(j,h)\right)$$
$$+V_f(i,h)\sum_{\substack{j=1\\j\neq i}}^{N}\left(G(i,j)V_f(j,h)+B(i,j)V_e(j,h)\right)+P(i,h)=0, \quad i\in N \quad (3.23)$$

where G and B are the real and imaginary components of the complex admittance matrix elements, respectively, V_e and V_f are the real and imaginary components of the complex voltage, respectively, N is the total number of buses, i, j, are indices for buses. The active power injection at bus i during hour h is calculated by

$$P(i,h) = P_d(i,h) - P_r(i,h)\beta_{c.r}(i,h) - P_g(i,h)$$
$$- P_{dis}(i,h) + P_{ch}(i,h), \quad i \in N \quad (3.24)$$

where $P_r(i,h)$ is the active power of a RGE at bus i during hour h, $\beta_{c.r}(i,h)$ is the curtailment factor of a RGE at bus i during hour h and $P_{ch}(i,h)$, $P_{dis}(i,h)$ are the active power charge and discharge of a BSS i during hour h, respectively. The reactive power flow equation is given by

$$V_f(i,h)\sum_{\substack{j=1\\j\neq i}}^{N}\left(G(i,j)V_e(j,h)-B(i,j)V_f(j,h)\right)$$
$$-V_e(i,h)\sum_{\substack{j=1\\j\neq i}}^{N}\left(G(i,j)V_f(j,h)+B(i,j)V_e(j,h)\right)+Q(i,h)=0, \quad i\in N \quad (3.25)$$

where the reactive power injection at bus i during hour h is calculated by

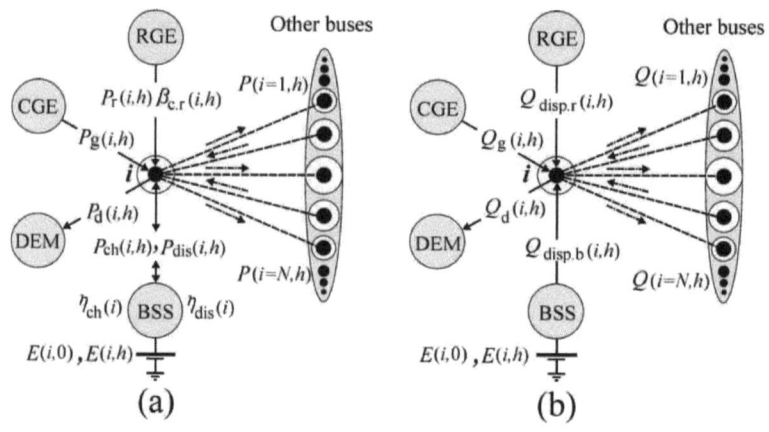

Figure 3.14: Power and energy balance at bus i: (a) active power (b) reactive power.

$$Q(i,h) = Q_d(i,h) - Q_g(i,h) - Q_{disp.r}(i,h) - Q_{disp.b}(i,h), \quad i \in N \quad (3.26)$$

where $Q_{disp.r}(i,h)$ is the reactive power of a RGE at bus i during hour h and $Q_{disp.b}(i,h)$ is the reactive power of a BSS at bus i during hour h. The hourly energy balance in each BSS at bus i during hour h can be written as

$$E(i,h+1) - E(i,h) - \eta_{ch} P_{ch}(i,h) + P_{dis}(i,h)/\eta_{dis} = 0, \quad i \in N \quad (3.27)$$

$$E(i, T_{initial}) = E(i, T_{final}), \quad i \in N. \quad (3.28)$$

It is to note that reactive power dispatches cause losses in the B-PCSs. However, in this work, we neglect this loss, since its effect on the overall benefit is much less than the reactive power which will cover the reactive energy demand in DNs which will be used to minimize grid losses, minimize the reactive energy import from an upstream connecting network, and improve voltage profiles. This is also valid for RGEs. Briefly, RGEs and BSSs work (when producing/absorbing reactive power) as ideal capacitor banks or inductors.

3.3.5 Definition of Infinite Bus

As described in Section (3.1), in a PDN unidirectional power flows are expected, i.e., both active and reactive power need to be imported from an upstream connecting network at the secondary bus S_1 of a main TR (see Fig. 3.1). This leads to active and reactive power shapes at this bus in accordance with the demand shown in Fig. 3.5.

However, in an ADN bidirectional power flows for both active and reactive power can be observed. In other words, active and/or reactive power can be imported in the forward direction (from bus S_0 to bus S_1) or exported (from bus S_1 to bus S_0), as seen in Fig. 3.2. Consequently, the shapes of active and reactive power profiles at bus S_1 can have complex shapes, as seen in Fig. 3.15.

In this work, bounds are considered on the forward and reverse power flows at slack bus S_1 as follows

$$S_{S1}^2(h) = P_{S1}^2(h) + Q_{S1}^2(h), \tag{3.29}$$

$$S_{S1}(h) \leq S_{S1.\max}, \tag{3.30}$$

$$-\alpha_{P1.rev}\, S_{S1.\max} \leq P_{S1}(h) \leq \alpha_{P1.fw}\, S_{S1.\max}, \tag{3.31}$$

$$-\alpha_{Q1.rev}\, S_{S1.\max} \leq Q_{S1}(h) \leq \alpha_{Q1.fw}\, S_{S1.\max}, \tag{3.32}$$

$$0 \leq \alpha_{P1.fw} \leq 1, \tag{3.33}$$

$$0 \leq \alpha_{Q1.fw} \leq 1, \tag{3.34}$$

$$0 \leq \alpha_{P1.rev} \leq 1, \tag{3.35}$$

$$0 \leq \alpha_{Q1.rev} \leq 1, \tag{3.36}$$

where $S_{S1}(h)$, $P_{S1}(h)$ and $Q_{S1}(h)$ are apparent, active and reactive power, respectively, at slack bus during hour h, $S_{S1.\max}$ is the upper bound of apparent power at slack bus, $\alpha_{P1.fw}$, $\alpha_{Q1.fw}$ are upper bounds of active and reactive power in forward direction at slack bus, respectively, and $\alpha_{P1.rev}$, $\alpha_{Q1.rev}$ are upper bounds of active and reactive power in reverse direction at

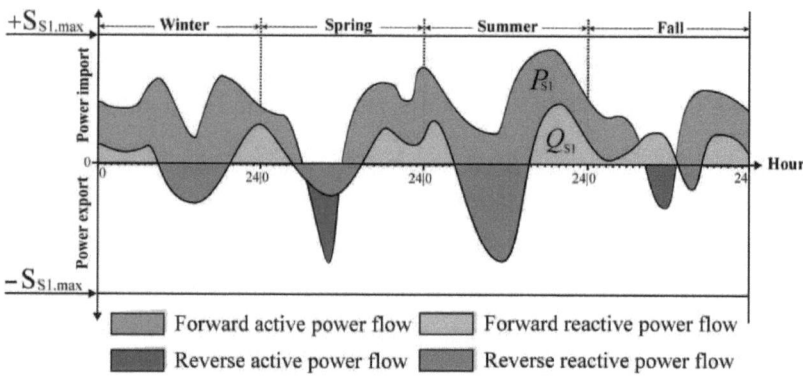

Figure 3.15: Conceptual daily active and reactive power exchange at the slack/infinite bus in an ADN [40].

slack bus, respectively. Typically, the power flow in DNs in forward direction is "normal", but because of a high penetration of embedded DG units in DNs upper bounds in reverse direction should be consider for active [7] and reactive power [40].

3.4 Modeling of Energy Prices

3.4.1 Forward Active-Reactive Energy Prices

In modern power systems generation companies produce electricity at "relatively" low prices while transmission and distribution companies take the task to transmit/deliver it away to final consumers. This process leads to: 1) higher electricity prices towards final consumers and 2) additional demand in form of losses in transmission and distribution networks (TNs/DNs). Note that in this work ideal TRs [36] (i.e., with neglected losses) are considered, as seen in Fig. 3.16. In addition, only average active [121] and reactive [42] energy prices are depicted.

It is noted that reactive energy prices are applicable in different countries with different forms. In common practice, the cost of transporting the reactive energy is added to final consumer's bills as: 1) cost of active energy losses made in both TNs and DNs and 2) cost of reactive energy transport through both TNs and DNs.

3.4.2 Feed-in-Tariffs and Reverse Active Power Flow

Generally, governmental regulations stand behind the remuneration of REGs (such as wind and PV) [41]. This leads to so-called FIT systems. It is to note that FIT systems are being applied differently in different countries (e.g., Germany and Ontario [71]). It is worth mentioning that considering a high penetration of REGs at a specific voltage level will lead to bidirectional power flows, i.e., reverse active energy. In addition, the output power of REGs will be curtailed due to system constraints. Therefore, all these energies and its prices at all voltage levels should be considered, as depicted in Fig. 3.17. Here, we use dots for representing different energy prices, which can be used in the future, for charging and remunerating different types of energies (i.e., demand, losses, renewable active energy, reverse active energy, and curtailed active energy) at different voltage levels (i.e., LV, MV, HV, and VHV).

3.4.3 Charge-Remuneration Rates for Battery Storage

In future grids, BSSs are expected to be added in the power system [7][40]. This brings additional new energies (i.e., from active power charge and discharge of BSSs) which should be considered. This increases the complexity of energy prices, as shown in Fig. 3.18. Moreover, if the reactive power capability of these BSSs is utilized, then reverse reactive energies can be observed [40][42].

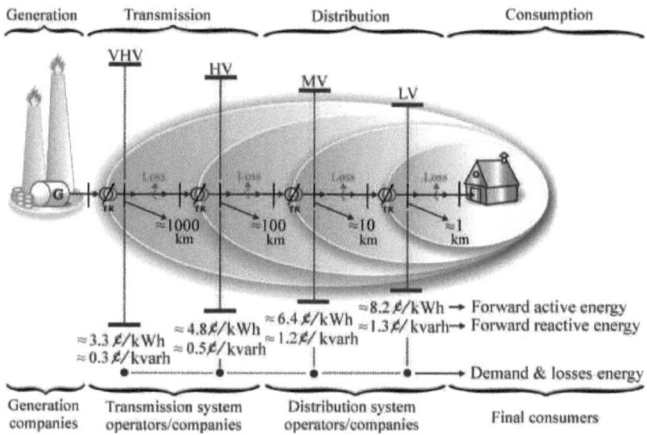

Figure 3.16: Conceptual illustration of active-reactive energy prices in the modern power system. Prices indicate the average cost/kWh [121] and cost/kvarh [42].

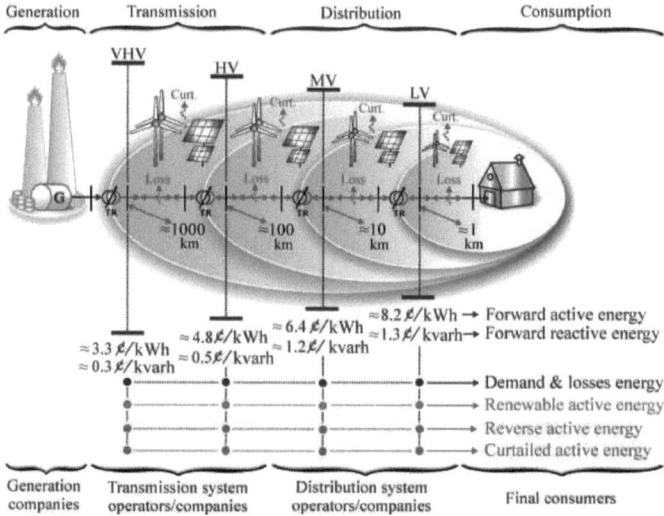

Figure 3.17: Conceptual illustration of active-reactive energy prices in the modern power system with REGs, curtailments and reverse active power flow.

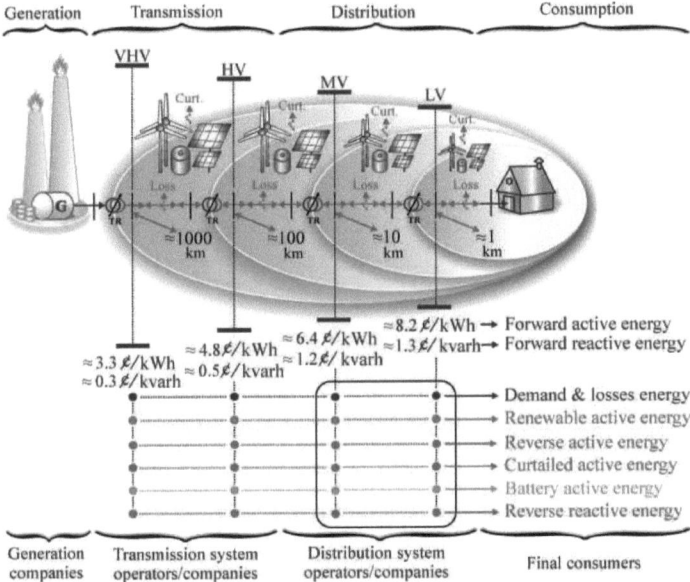

Figure 3.18: Conceptual illustration of active-reactive energy prices in the modern power system with REGs, curtailments, reverse active-reactive power flows, and BSSs.

In this work, for the sake of simplicity, we consider active and reactive energy prices in the DS as given in the following chapters. These prices are grouped in a rectangular as shown in Fig. 3.18. Here, we note that a meter-based method is proposed in this work to charge and remunerate different types of energies as explained in the following.

3.4.4 Meter-Based Method for Charging and Remunerating

It is expected that electrical meters or smart meters will be more and more integrated in power systems in the case of a high penetration of renewable energies, and therefore, a meter-based method is used in this work. Briefly, it is based on charging and remunerating all entities connected to a power

Chap. 3: Modeling Procedures

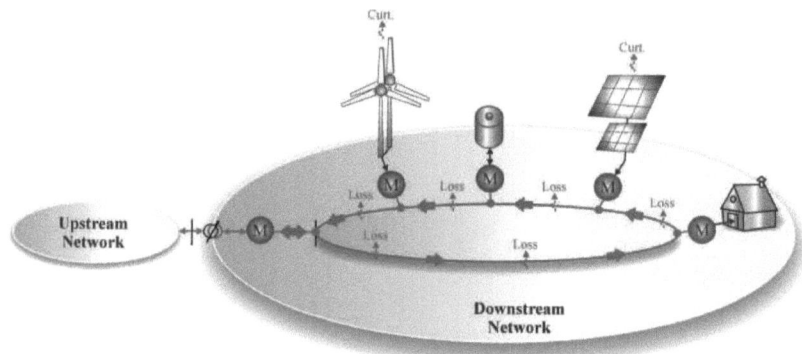

Figure 3.19: Conceptual illustration of the meter-based method for charging and remunerating different entities connected to a power system. Here, M stands for meter.

system based on the energy measured by the meters, as shown in Fig. 3.19. Note that power losses and curtailments, as seen in Fig. 3.19, are not measured directly using meters as used for other entities, i.e., wind and PV generators, BSSs, demand and forward/reverse power flows from/to an upstream connected network. However, they can be calculated mathematically as shown later.

4 Simulation and Optimization in Passive Distribution Networks

In this chapter, simulation procedures in PDNs (without DG units and BSSs) are made. This is made to examine the characteristics of PDNs (low- and medium-voltage networks) before considering any additional entities (such as REGs and/or BSSs). In particular, the effects of changing the voltage amplitude at the secondary bus S_1 (see Fig. 3.1) of main TRs used to connect downstream networks and upstream networks are explored. The simulation procedures are made using two DSIs which are developed and implemented in this work using the MATLAB-Simulink environment. The first DSI (called DSI-1) is used for carrying out dynamic power flow studies in DNs, whereas the second DSI (called DSI-2) is a control system of an OLTC transformer. Finally, to minimize active energy losses in PDNs an optimization framework derived from the A-R-OPF method [40] [42] is proposed.

4.1 Simulation in PDNs

4.1.1 Dynamic Power Flow in PDNs

The dynamic power flow studies in PDNs are carried out using the DSI-1, as seen in Appendix D. The flowchart of the dynamic power flow algorithm is shown in Fig. 4.1. Here, system parameters and demand power profiles of a PDN are given as input to the solver which is based on the Newton-Raphson method as given in Section (3.2.5). Briefly, at each time step h both active and reactive power flow equations are solved repeatedly until a convergence criterion is satisfied. Two criteria are used to terminate the computation process, i.e., either a norm of the correction vector $\Delta \mathbf{X}$ is less than or equal to a predefined small tolerance, e.g., $\varepsilon = 10^{-9}$ or a maximum number of iterations, e.g., $it_{max} = 100$ is reached. It is noted that many factors affect the convergence process, such as system parameters and demand profiles, as illustrated in the case studies below. The obtained solution from the solver is saved and the computation process is repeated for the next time step until T_{final}, as seen in Fig. 4.1. Note that generation power profiles are assigned to zeros for PDNs.

4.1.2 Control System of an OLTC Transformer

In this work, we use the OLTC transformer control system as in [15] for analysis purposes. The block diagram of this system is shown in Fig. 4.2 and implemented in the MATALAB-Simulink environment as given in Appendix E. The goal of using such a control system is to hold the voltage amplitude at a slack bus near (as possible) to pre- defined set points $|V_{S1.ref}(h)|$ by changing the transformer tap ratio a(h). In Fig. 4.2, PT(h) represents active power TR load during time step h, QT(h) represents reactive power TR load during time step h and $V_{S0}(h)$ represents primary voltage of a TR during time step h. Note that a time step is based on the discretization of a time horizon. It can be in hour, minute or second, as shown in the case studies below.

4.1.3 Case Studies

In this section, parameters and demand power profiles of two real DNs are used and simulation results are presented. The first DN is a low-voltage DN, as seen in Fig. 4.3, while the second DN is a medium-voltage DN, as seen in Fig. 4.4. The IEEE-RTS load data as a percentage of the annual peak load is considered for both networks (see Appendix A).

Chap. 4: Simulation and Optimization in Passive Distribution Networks

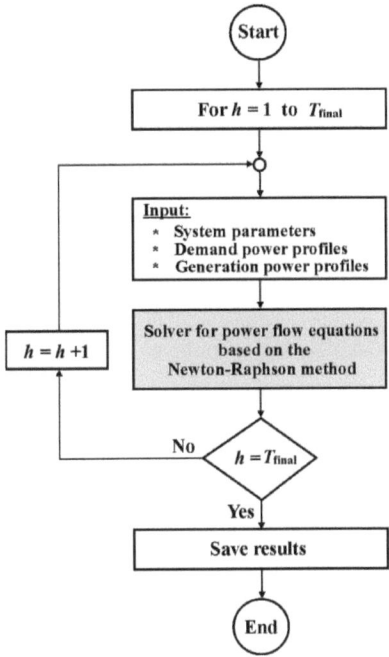

Figure 4.1: Flowchart of the dynamic power flow algorithm (DSI-1).

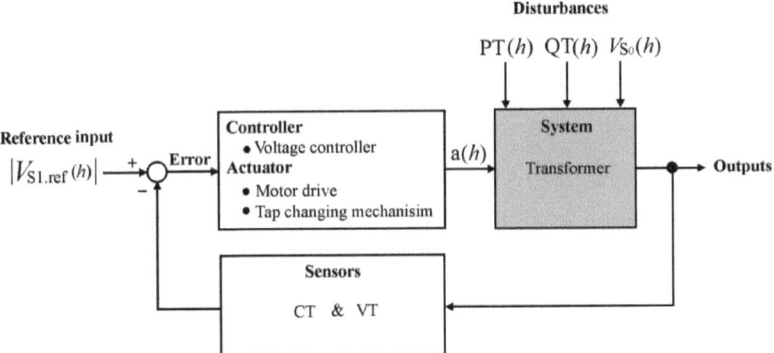

Figure 4.2: Block diagram of an OLTC control system (DSI-2).

In addition, the bus number 1 is selected as slack bus S_1 (fixed voltage amplitude and 0 phase angle), whereas the rest are considered as PQ buses.

The low-voltage DN is a real three phase balanced DN. This network was studied in [4][12]. It contains 29-buses at 0.4 kV connected to a medium-voltage bus S_0. The maximum feeder capacity is taken 75 kVA. The data of this network is given in Appendix B: Table B.1 (network data) and Table B.2 (demand peak and PFs). For this network, values in per unit system are given on 100-kVA base, unless otherwise specified. The medium-voltage DN is also a real three phase balanced radial DN. This network was studied in [5][6][7][8]. It contains 41-buses at 27.6 kV connected to a high-voltage bus S_0. The maximum feeder capacity is taken 14.3 MVA. The data of this network is given in Appendix C: Table C.1 (network data) and Table C.2 (demand peak and PFs). Again, for this network, values in per unit system are given on 10-MVA base, unless otherwise specified.

The simulation results of these networks are drawn in Figs. 4.5 and 4.6 (for the low-voltage DN), while in Figs. 4.7 and 4.8 (for the medium-voltage DN). Here, simulations are done in four typical season's days with one hour time step, i.e., 4 × 24 = 96 hour. Fig. 4.5(a) and Fig. 4.7(a) show active and reactive power demand for the low- and medium-voltage network, respectively. These demand profiles are given as input to the DSI-1. It is clearly seen from the curves, shown in Figs. 4.5(b) and 4.7(b), that voltage amplitudes are being violated during typical season's days, especially during heavy demand in summer days. Note that the specified voltage amplitudes at slack buses are set to 1 pu. These specifications have an important role in terms of voltage violation. For example, if the voltage amplitude at a slack bus is set to a value higher than 1 pu, the voltage violation at the lower bound can be avoided at PQ buses. Note that the voltage angles at PQ buses in the low-voltage DN are positive, as seen in Fig. 4.5(c), whereas they are negative in Fig. 4.7(c). This fact comes from the difference in network parameters in both low- and medium-voltage networks, especially the ratio of resistance R_1 and inductive reactance X_1 of line segments. Figs. 4.5(d) and 4.7(d) show the active and reactive power at slack bus in the low- and medium-voltage DN, respectively. These values are higher than the demand in both DNs because of active and reactive energy losses in the grids, as seen in Figs. 4.6(a) and 4.8(a).

From another perspective, it can be clearly seen that not more than five iterations are needed for the DSI-1 to converge, as seen in Figs. 4.6(b) and 4.8(b), even with different condition numbers of the Jacobian matrixes, as seen in Figs. 4.6(c) and 4.8(c). The predefined small tolerance set for both cases is $\varepsilon = 10^{-9}$. This value is satisfied as shown in Figs. 4.6(d) and 4.8(d). It is interesting to see that the condition numbers of the Jacobian matrixes and number of iterations are in accordance with the demand.

In the above scenarios, the voltage amplitude at slack buses is assumed to be fixed with 1 pu in the time horizon. Practically, this assumption can be realized by using an OLTC transformer control system as explained next.

The DSI-2 is used to carry out simulations for both low- and medium-voltage DNs. Here, simulations are done in four typical season's days with one second time step, i.e., $4 \times 24 \times 60 \times 60 = 345\ 600$ second. Based on [15], the system parameters used in the simulation are given as follows:
- The peak transformer load $PT_{peak} + j\ QT_{peak}$ is taken $(47 + j\ 22.76)$ kVA for the low-voltage DN, while $(15.2 + j\ 5.52)$ MVA for the medium-voltage DN. This peak load is multiplied by the hourly IEEE-RTS load

data (see Appendix A), and random numbers generated from the normal distribution with a mean parameter equals 1 and a standard deviation parameter equals 0.02. Note that these random numbers are generated every 30 minute, i.e., every 1800 second.

- The mean value of the primary transformer voltage is taken 1 pu for the low-voltage and medium-voltage DN. This base voltage is multiplied by random numbers generated from the normal distribution with a mean parameter 1 and a standard deviation parameter 0.008. Note that these random numbers are generated every 30 minute, i.e., every 1800 second.
- Both transformer and compensating impedances are neglected in this work.
- The reference voltage is taken 1.05 pu for the low-voltage DN, while 1 pu for the medium-voltage DN.
- The tap range is taken ±10 taps with a regulation step equals 0.01 pu. The initial tap position is taken 0 for all simulation procedures.
- An inverse-time constant is 100 second, while the dead-band is 0.01 pu.
- A motor drive delay time is 10 second.

From the assumed parameters above, we obtain profiles for the active and reactive power load and primary voltage of a TR, as seen in Figs. 4.9(a-b) and 4.11(a-b). These are considered disturbances for the system, as shown in Fig. 4.2. Different set points as references are used for the low-voltage DN (with $|V_{S1.ref}(h)| = 1.05$ pu) and the medium-voltage DN (with $|V_{S1.ref}(h)| = 1.00$ pu). It can be seen from Figs. 4.9(c) and 4.11(c) that the control system can hold the voltage at the slack bus as near as possible to the predefined set points based on the assumed dead-band as shown in Figs. 4.9(d) and 4.11(d). The tap position and transformer tap-ratio are also drawn in Figs. 4.10(a-b) and 4.12(a-b). It is obvious that the tap position is in accordance with the transformer tap-ratio which reflects the basic idea of operating an OLTC transformer [15]. Here, we conclude:

1) Using an OLTC of a main TR in a DN can hold the voltage at the secondary bus in an acceptable range. Consequently, the voltage at other PQ buses inside the DN can also be maintained in a predefined range.
2) The set points of voltage at a slack bus affect energy losses in the grid, e.g., a high value for the voltage at the slack bus leads to less energy losses.
3) A flexible optimization framework is needed to minimize the total energy losses in PDNs. This is investigated in the next section.

Chap. 4: Simulation and Optimization in Passive Distribution Networks

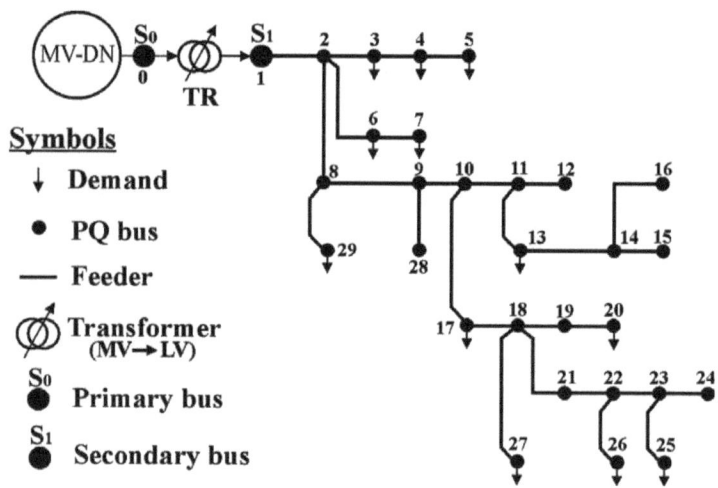

Figure 4.3: Low-voltage network for the case study as PDN.

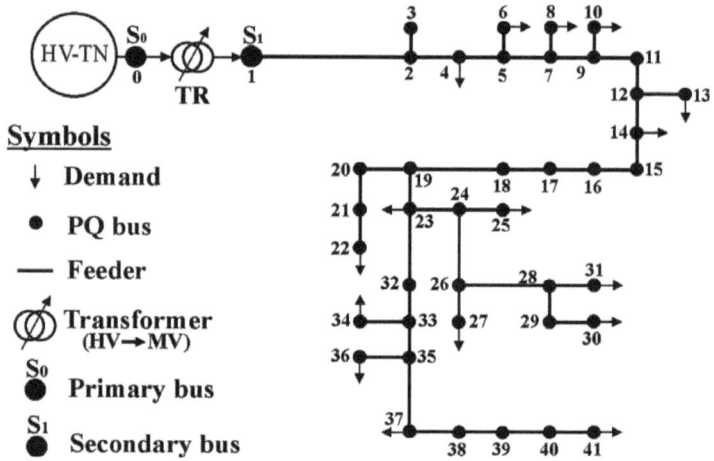

Figure 4.4: Medium-voltage network for the case study as PDN.

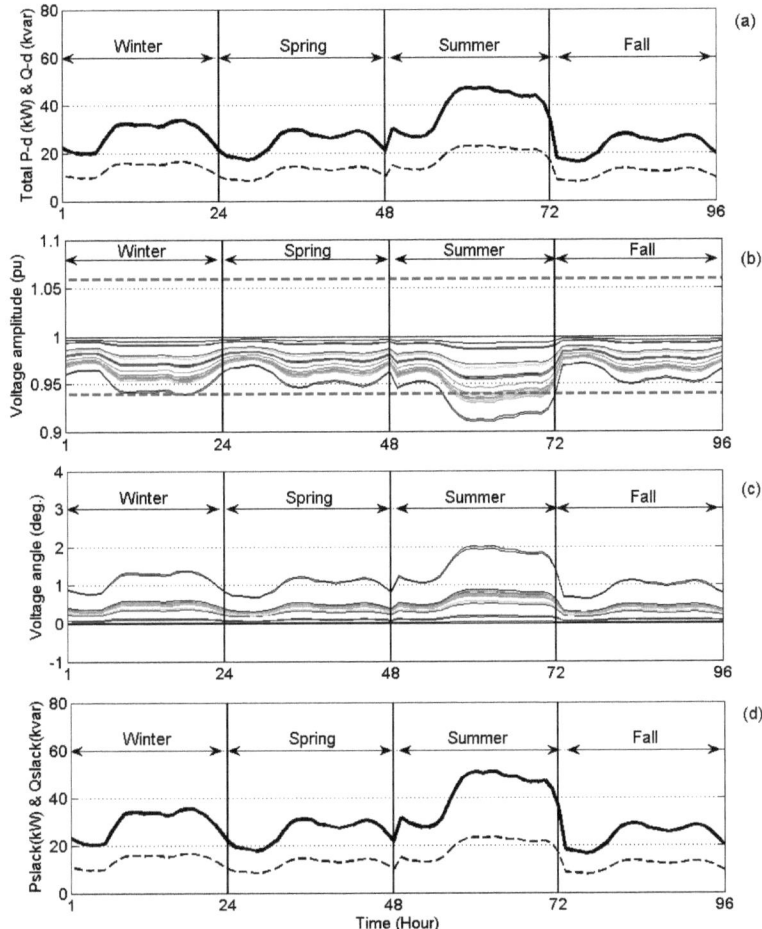

Figure 4.5: Simulation results of the low-voltage DN in four typical season's days: (a) Total demand active power (solid-bold) and reactive power (dashed-thin). (b) Voltage amplitude of all buses (c) Voltage angle of all buses. (d) Slack bus active power (solid-bold) and reactive power (dashed-thin).

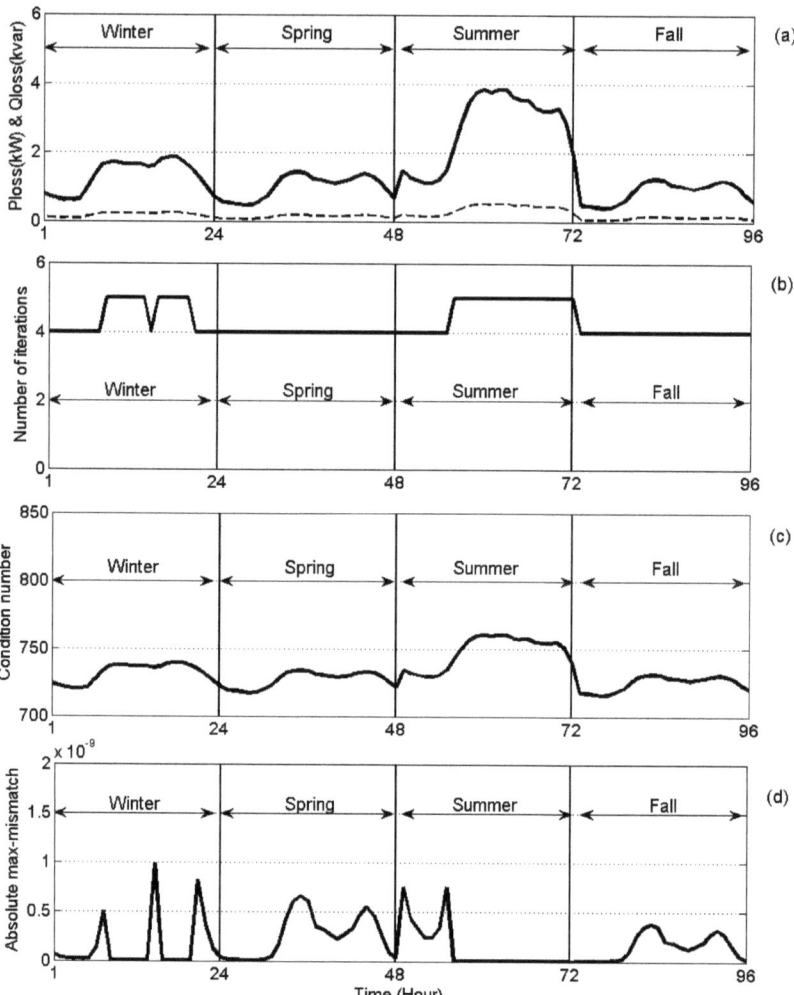

Figure 4.6: Simulation results of the low-voltage DN in four typical season's days: (a) Total active power losses (solid-bold) and reactive power losses (dashed-thin). (b) Number of iterations. (c) Condition number. (d) Absolute maximum mismatch.

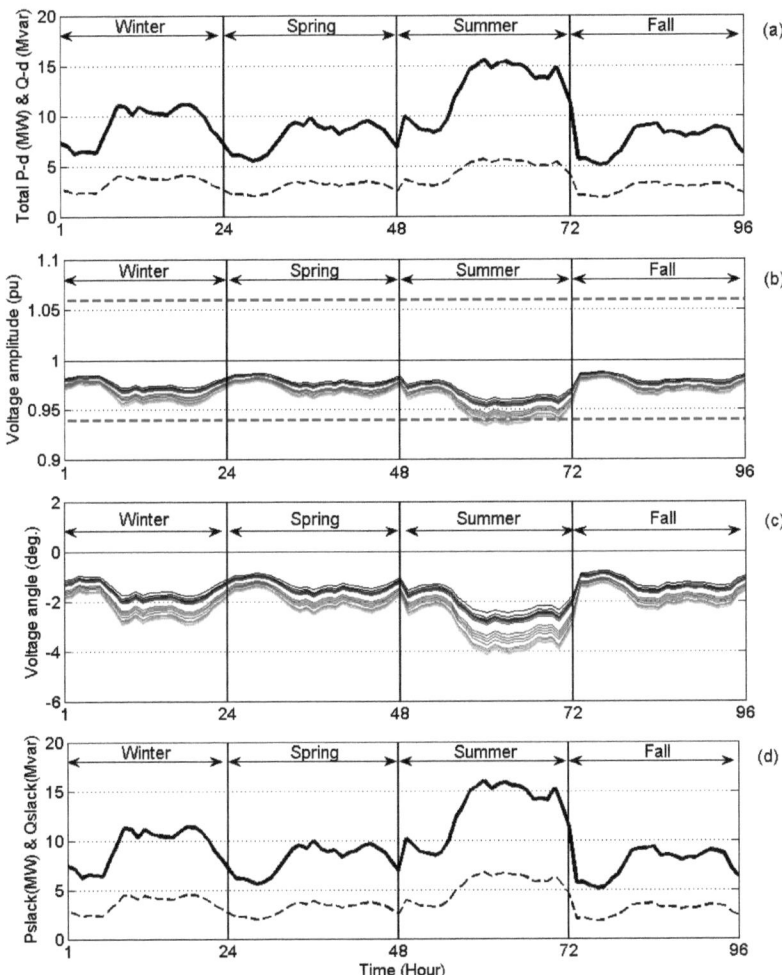

Figure 4.7: Simulation results of the medium-voltage DN in four typical season's days: (a) Total demand active power (solid-bold) and reactive power (dashed-thin). (b) Voltage amplitude of all buses (c) Voltage angle of all buses. (d) Slack bus active power (solid-bold) and reactive power (dashed-thin).

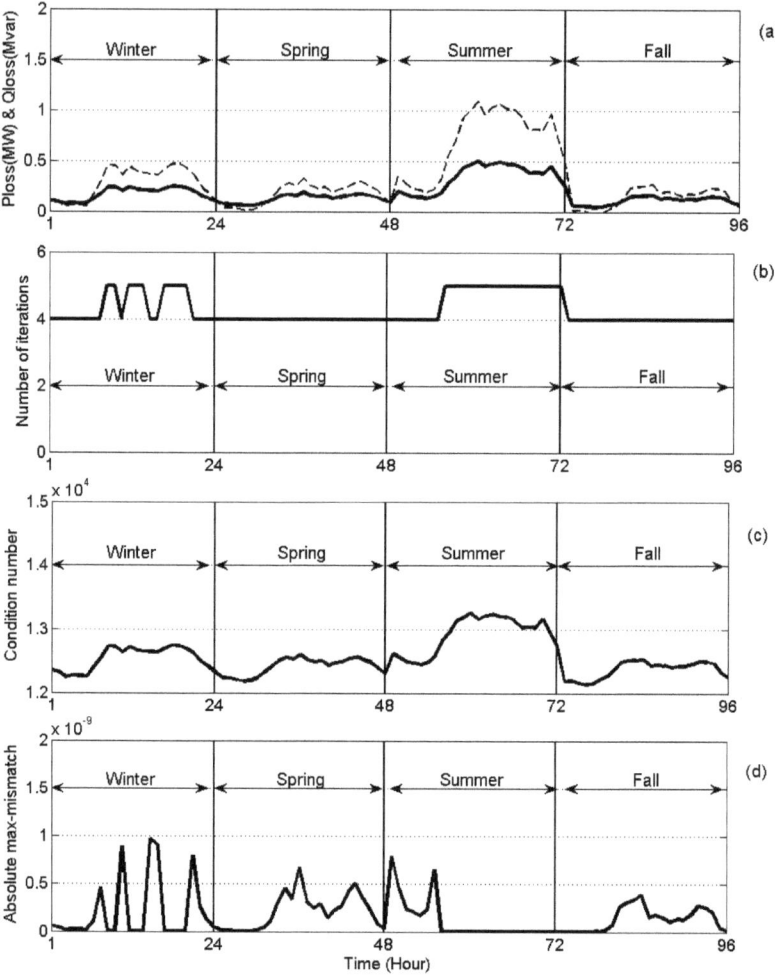

Figure 4.8: Simulation results of the medium-voltage DN in four typical season's days: (a) Total active power losses (solid-bold) and reactive power losses (dashed-thin). (b) Number of iterations. (c) Condition number. (d) Absolute maximum mismatch.

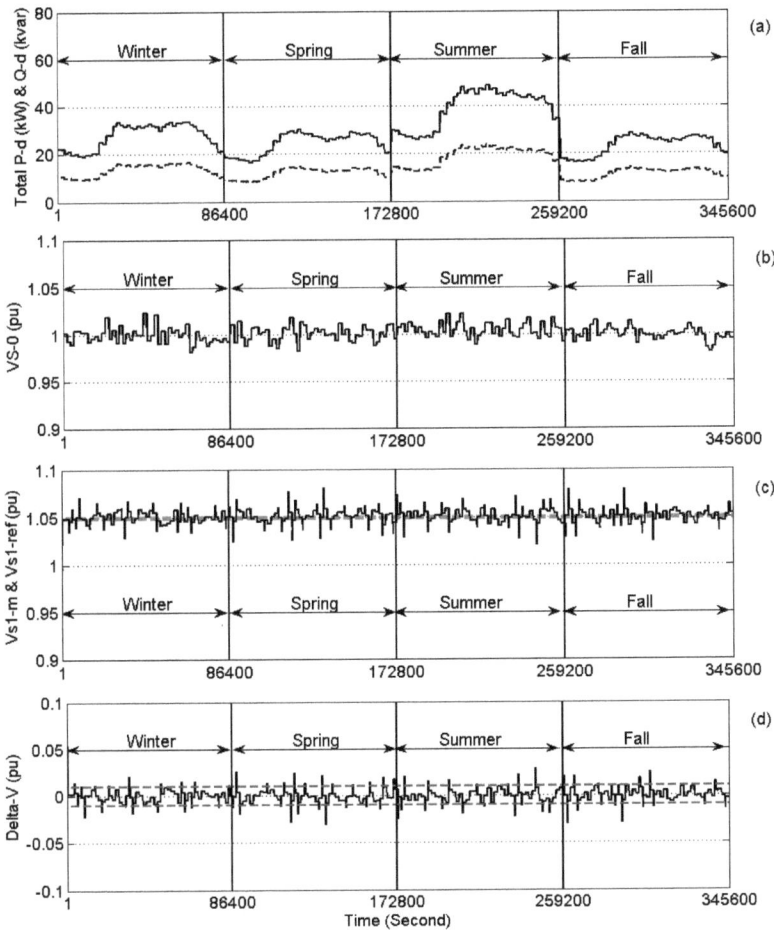

Figure 4.9: Simulation results of the low-voltage DN in four typical season's days: (a) Transformer active power load (solid) and reactive power load (dashed). (b) Primary voltage of a TR. (c) Measured secondary voltage of a TR (solid-black) and reference voltage (dashed-green). (d) Voltage error (solid-black) and dead-band (dashed-red).

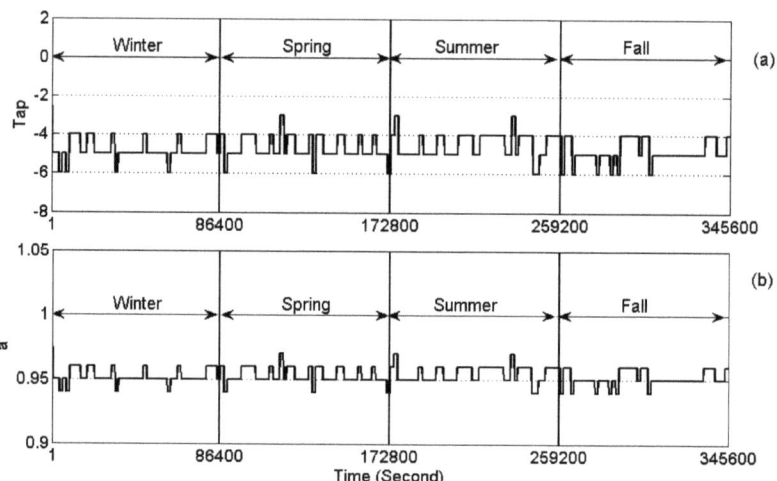

Figure 4.10: Simulation results of the low-voltage DN in four typical season's days: (a) Tap position. (b) Transformer tap-ratio.

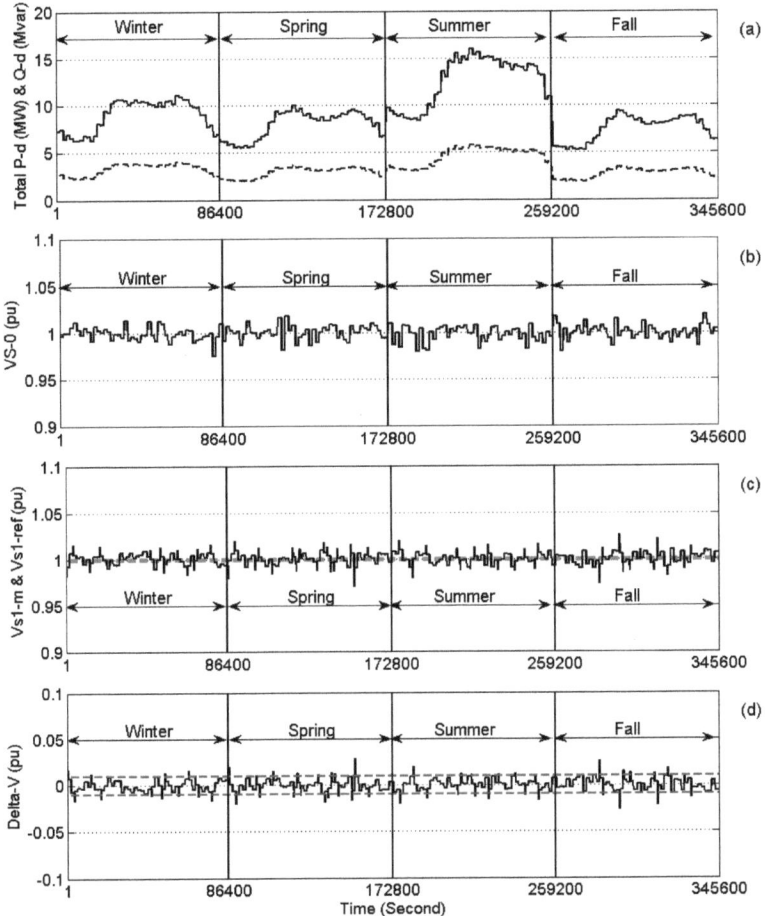

Figure 4.11: Simulation results of the medium-voltage DN in four typical season's days: (a) Transformer active power load (solid) and reactive power load (dashed). (b) Primary voltage of a TR. (c) Measured secondary voltage of a TR (solid-black) and reference voltage (dashed-green). (d) Voltage error (solid-black) and dead-band (dashed-red).

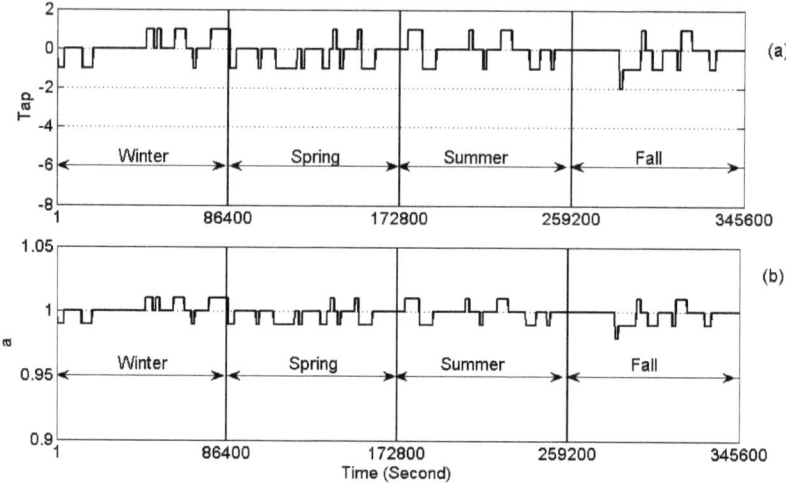

Figure 4.12: Simulation results of the medium-voltage DN in four typical season's days: (a) Tap position. (b) Transformer tap-ratio.

4.2 OPF in PDNs Utilizing OLTCs Capability

4.2.1 Optimal Voltage Regulation in PDNs

The objective function of voltage control in traditional PDNs (without DG units and BSSs) can be either to minimize feeder losses or to operate the feeder closed to the nominal voltage [115].

In the A-R-OPF method in low-voltage DNs [41] and medium-voltage DNs [40][42], an OLTC was considered with fixed set points for voltage amplitudes $|V_{S1.ref}(h)|$ at each hour h. It means that $|V_{S1.ref}(h)|$ at the secondary bus (slack bus S_1) of main TRs are considered constant in a time horizon. In this work, we investigate the properties of network operations when an OLTC is considered in the OPF problem.

In particular, we evaluate the active energy losses in PDNs when changing $|V_{S1.ref}(h)|$. This leads to an optimization problem in which $|V_{S1.ref}(h)|$ is the sole control variable. Typically, this control variable has a discrete nature as seen from the search space in Fig. 4.13. We formulate the following optimization problem in which the total active energy losses are minimized as follows:

$$\min_{|V_{S1.ref}|} Losses \qquad (4.1)$$

where

$$Losses = \frac{1}{2}\sum_{h=1}^{T_{final}}\sum_{i=1}^{N}\sum_{j=1}^{N} G(i,j)\left(\left(V_e(i,h)\right)^2 \right.$$
$$\left. +\left(V_f(i,h)\right)^2 +\left(V_e(j,h)\right)^2 +\left(V_f(j,h)\right)^2 \right.$$
$$\left. -2\left(V_e(i,h)V_e(j,h)+V_f(i,h)V_f(j,h)\right) \right) \qquad (4.2)$$

subject to the equality and inequality constraints that include the active power flow equations at the buses as follows:

$$V_e(i,h) \sum_{\substack{j=1 \\ j\neq i}}^{N}\left(G(i,j) V_e(j,h) - B(i,j) V_f(j,h)\right)$$
$$+V_f(i,h) \sum_{\substack{j=1 \\ j\neq i}}^{N}\left(G(i,j) V_f(j,h) + B(i,j) V_e(j,h)\right) + P(i,h) = 0, \quad i \in N \qquad (4.3)$$

where the active power injection at bus i during hour h is calculated by

$$P(i,h) = P_d(i,h) - P_{S1}(h), \qquad i \in N. \qquad (4.4)$$

The reactive power flow equation is given by

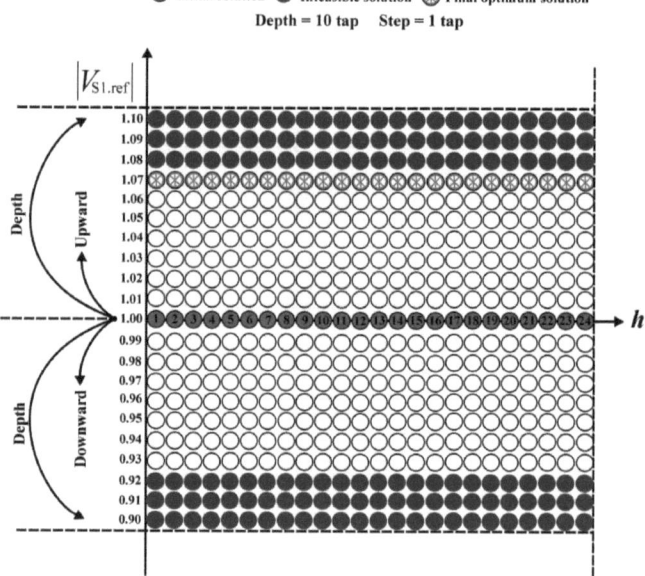

Figure 4.13: Illustration of the search space.

$$V_f(i,h) \sum_{\substack{j=1 \\ j\in i}}^{N} \bigl(G(i,j)\, V_e(j,h) - B(i,j)\, V_f(j,h)\bigr)$$

$$-V_e(i,h) \sum_{\substack{j=1 \\ j\in i}}^{N} \bigl(G(i,j)\, V_f(j,h) + B(i,j)\, V_e(j,h)\bigr) + Q(i,h) = 0, \quad i \in N \quad (4.5)$$

where the reactive power injection at bus i during hour h is calculated by

$$Q(i,h) = Q_d(i,h) - Q_{S1}(h), \qquad i \in N. \quad (4.6)$$

The inequality constraints consist of the satisfaction of voltage bounds

$$V_{\min}(i) \le V(i,h) \le V_{\max}(i) \qquad i \in N(i \ne S1), \quad (4.7)$$

active and reactive bounds at the slack bus (see Eqs. (3.30), (3.31) and (3.32)) and the main feeder bounds

$$S(i,j,h) \le S_{l.\max}(i,j), \qquad i,j \in N(i \ne j). \tag{4.8}$$

4.2.2 Proposed Method

The basic idea of the proposed method here is derived from [42], where a two level optimization framework was presented. The input-output scheme of this optimization framework is depicted in Fig. 4.14. Here the upper level is implemented to optimize the discrete control variable $|V_{S1.ref}(h)|$, and then it is delivered to the lower stage. In the lower stage the OPF problem described above is solved by a NLP solver. The resulting value of the objective function (Losses) and its state are brought to the upper stage for the next iteration. Here, a state of a solution means either feasible (i.e., state = 1) or infeasible (i.e., state = 0). The proposed algorithm can be summarized by the following steps:

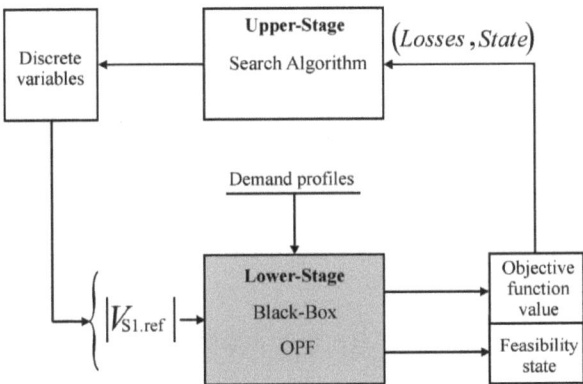

Figure 4.14: Input-output scheme for the OPF in PDNs with a search algorithm.

1) Choose an initial voltage amplitude $|V_{S1.ref}(h)|$ as a string for all hours in a day (e.g., the red string in Fig. 4.13 with $|V_{S1.ref}(h)| = 1$ pu).

2) Provide this initial string to the lower stage to evaluate the objective function value and its state. Then record this fitness in a register.

3) Sweep $|V_{S1.ref}(h)|$ by changing a tap (the search step is taken to be one tap) in the upward as well as downward direction, as shown in Fig. 4.13, with a specific depth (10 tap which is "applicable" in practice).

4) Sort the fitness of the successive evaluated strings in ascending order and retain the feasible and minimum of them (if two or more strings have the same fitness, the algorithm preserves the original ordering of the fitnesses). The total number of evaluations needed to find a converged solution is 21, given the initial $|V_{S1.ref}(h)| = 1$ pu.

5) The string, of all feasible solutions, with the minimum value of the objective function represents the optimal operation for a specific day.

4.2.3 A Case Study

The medium-voltage DN in Fig. 4.4 is used here to show that effectiveness of the proposed method. The results of solving the optimization problem shown in Fig. 4.14 are summarized in Table 4.1. It is clearly seen that a higher reduction of energy losses can be achieved if the capability of an OLTC is fully utilized. Here, fixed-OLTC means that $|V_{S1.ref}(h)| = 1$ pu, while flexible-OLTC means that $|V_{S1.ref}(h)|$ is free in a range (0.90-1.10 pu), as seen in Fig. 4.13.

It is worth mentioning here that other search algorithms are also possible to explore the search space dynamically. It means that $|V_{S1.ref}(h)|$ can be different from hour to hour. Of course this increases the complexity of the problem, but using more efficient computational frameworks, such as parallel computing, can improve the computational performance.

Table 4.1: Total losses for fixed and flexible set points of an OLTC
(One winter day for the medium-voltage PDN)

Total losses (MWh/1-day) Fixed-OLTC	Total losses (MWh/1-day) Flexible-OLTC	Difference
3.837(1.00)	Infeasible (1.10)	---
3.837(1.00)	Infeasible (1.09)	---
3.837(1.00)	Infeasible (1.08)	---
3.837(1.00)	3.317 (1.07)	0.52(**13.55%**)
3.837(1.00)	3.385 (1.06)	0.452(**11.70%**)
3.837(1.00)	3.454 (1.05)	0,383(**9.98%**)
3.837(1.00)	3.526 (1.04)	0.311(**8.10%**)
3.837(1.00)	3.600 (1.03)	0.237(**6.17%**)
3.837(1.00)	3.677 (1.02)	0.16(**4.16%**)
3.837(1.00)	3.756 (1.01)	0.081(**2.11%**)
3.837(1.00)	3.837 (1.00)	0(**0%**)
3.837(1.00)	3.922 (0.99)	-0.085(**-2.21%**)
3.837(1.00)	Infeasible (0.98)	---
3.837(1.00)	Infeasible (0.97)	---
3.837(1.00)	Infeasible (0.96)	---
3.837(1.00)	Infeasible (0.95)	---
3.837(1.00)	Infeasible (0.94)	---
3.837(1.00)	Infeasible (0.93)	---
3.837(1.00)	Infeasible (0.92)	---
3.837(1.00)	Infeasible (0.91)	---
3.837(1.00)	Infeasible (0.90)	---

5 Active-Reactive Optimal Power Flow in Active Distribution Networks

In this chapter, new mathematical models and problem formulations for A-R-OPF in low-voltage and medium-voltage ADNs (with DG units and BSSs) are presented. On the low-voltage level, a high penetration of PVSs is considered in the network in order to reveal the impact of such a scenario. In particular, the reactive power capability of the inverters of these PVSs is explored. The total revenue from PVSs is maximized and meanwhile the total cost of energy losses and demand is minimized. On the medium-voltage level, a DN with a high penetration of wind energy and BSSs is considered. In this case, the total revenue from wind parks and BSSs is maximized and meanwhile the total cost of energy losses is minimized. It is found that a huge reduction in energy losses and reactive energy imports can be achieved.

5.1 A-R-OPF for Low-Voltage ADNs

5.1.1 Modeling of Network Demand, Generation and Energy Prices

The load model is taken as explained in Section 3.2.1, whereas the PV power penetration as given in Section 3.3.2. Fig. 5.1 describes two main operation conditions of the low-voltage ADN, where $P_{S1}^{(1)}$ represents active power at slack bus in cloudy days or no generation of PVSs, whereas $P_{S1}^{(2)}$ represents active power at slack bus in heavily sunny days. We assume that it is possible to transport energy produced by PVSs to an upper medium-voltage DN, as seen from the negative part. The price model is considered as follows.

Generally, governmental regulations stand behind the remuneration of renewable energy sources [41]. In this section, two types of prices are used for the analysis, as shown in Fig. 5.1. The first is an average remuneration tariff for active power ($C_{pr.art.p}$), and it usually called as a fixed FIT. This price is used mainly to remunerate PVSs during an investment period. Second, on-peak/off-peak price model ($C_{pr.p}$) which is used for charging the demand and losses in a specific utility. It is noted that $C_{pr.art.p} > C_{pr.p}$ since renewable energies are being supported by governmental regulations [41]. Moreover, this governmental support usually decreases from year to year with a degression rate which can vary between 1.5% and 21% per year for PV installations attached to or on top of buildings in Germany from 1 January 2012 [41]. Not that reactive energy prices are neglected in this case study.

5.1.2 A-R-OPF Utilizing PV-DG Reactive Power Capability

In this section, an A-R-OPF problem is formulated as a dynamic optimization problem for low-voltage ADNs with a high penetration of PVSs. The optimization framework can be described by Fig. 5.2. Here, we consider that the network is being operated by a distribution system operator (DSO) who is responsible for operating the network with a high quality. The DSO tends to maximize the total revenue from the PVSs and meanwhile minimize the total cost of losses and demand.

As shown in Fig. 5.2, the control variables are the curtailment factor for active power and the reactive power dispatch of PVSs. Both of these controls are required as ancillary services. The optimization problem is defined in the following.

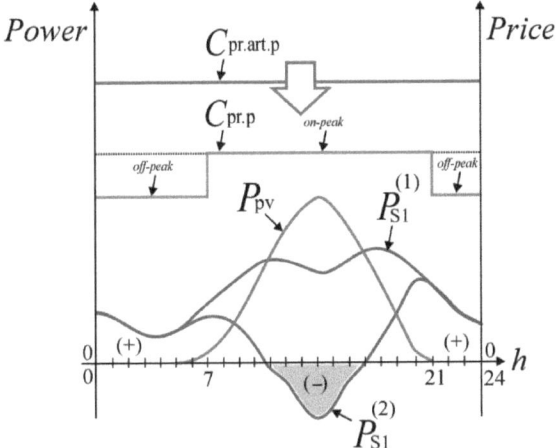

Figure 5.1: Daily photovoltaic/slack bus power profiles and energy prices. Here, $C_{pr.p}$ stands for a two-tariff price model of active energy and $C_{pr.art.p}$ for an average remuneration tariff price model.

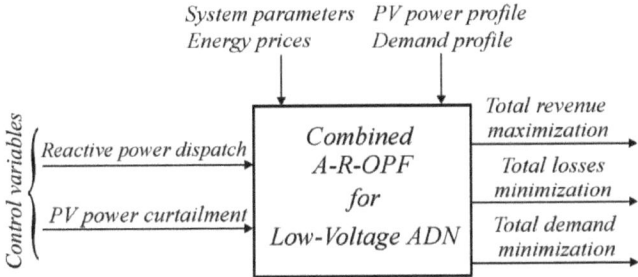

Figure 5.2: Input-output scheme for the combined A-R-OPF for the low-voltage ADN.

Objective function:

The aim of optimization is defined to maximize a multi-criteria objective function. It includes the total revenue from all PVSs connected to the low-voltage ADN, meanwhile the total cost will be minimized involving the cost of power losses and the cost of demand, i.e.,

$$\max_{Q_{disp.pv}, \beta_{c.pv}} F_{pv} - F_{loss} - F_{demand}, \quad (5.1)$$

where the three terms include the total revenue of the PVSs

$$F_{pv} = \sum_{h=1}^{T_{final}} C_{pr.p.art}(h) \sum_{\substack{i=1 \\ i \in Ipv}}^{N} P_{pv}(i,h) \beta_{c.pv}(i,h), \quad (5.2)$$

the loss cost

$$F_{loss} = \frac{1}{2} \sum_{h=1}^{T_{final}} C_{pr.p}(h) \sum_{i=1}^{N} \sum_{j=1}^{N} G(i,j) \Big(\big(V_e(i,h)\big)^2$$
$$+ \big(V_f(i,h)\big)^2 + \big(V_e(j,h)\big)^2 + \big(V_f(j,h)\big)^2 \quad (5.3)$$
$$- 2\big(V_e(i,h) V_e(j,h) + V_f(i,h) V_f(j,h)\big) \Big),$$

and the demand cost

$$F_{demand} = \sum_{h=1}^{T_{final}} C_{pr.p}(h) \sum_{i=1}^{N} P_d(i,h). \quad (5.4)$$

Here, G is the real component of the complex admittance matrix elements, V_e, V_f are the real and imaginary components of the complex voltage, respectively, $P_d(i,h)$ is the demand at bus i during hour h, N is the total number of buses, i and j are indices for buses, Ipv is the set of PVSs, T_{final} is the final time point in the optimization horizon (one year, i.e., $T_{final} = 8760$ h). Note that the optimization problem is solved daily and repeated for one year. This is typically used for a short-term analysis in power systems. Briefly, the considered market strategy for the low-voltage ADN can be summarized as follows:

- The PV-based DG units and reverse active energy to the upstream medium-voltage DN are paid for the same price model, i.e., the average remuneration tariff.
- The active energy losses and demands are charged by the same price model, i.e., the on-peak/off-peak price model.

Equality constraints:

The equality constraints are the active power balance at each bus in the network as follows

$$
\begin{aligned}
&V_e(i,h) \sum_{\substack{j=1\\j\in i}}^{N} \big(G(i,j)\, V_e(j,h) - B(i,j)\, V_f(j,h)\big) \\
&+ V_f(i,h) \sum_{\substack{j=1\\j\in i}}^{N} \big(G(i,j)\, V_f(j,h) + B(i,j)\, V_e(j,h)\big) + P(i,h) = 0, \quad i \in N
\end{aligned} \quad (5.5)
$$

where B is the imaginary component of the complex admittance matrix elements and P is the active power injection which is given by

$$ P(i,h) = P_d(i,h) - P_{pv}(i,h)\beta_{c.pv}(i,h) - P_{S1}(h), \quad i \in N \quad (5.6) $$

where P_{S1} is the active power at the slack bus. The reactive power balance at each bus is given as follows

$$
\begin{aligned}
&V_f(i,h) \sum_{\substack{j=1\\j\in i}}^{N} \big(G(i,j)\, V_e(j,h) - B(i,j)\, V_f(j,h)\big) \\
&- V_e(i,h) \sum_{\substack{j=1\\j\in i}}^{N} \big(G(i,j)\, V_f(j,h) + B(i,j)\, V_e(j,h)\big) + Q(i,h) = 0, \quad i \in N
\end{aligned} \quad (5.7)
$$

where Q is the reactive power injection which is given by

$$ Q(i,h) = Q_d(i,h) - Q_{disp.pv}(i,h) - Q_{S1}(h), \quad i \in N \quad (5.8) $$

where Q_{S1} is the reactive power at the slack bus. Note that to show the impact of PVSs alone, BSSs are not considered here.

Inequality constraints:

The inequality constraints include the restrictions on PVSs (Eqs. (3.10) and (3.11)) as well as voltage bounds at each PQ bus

$$V_{\min}(i) \leq V(i,h) \leq V_{\max}(i), \quad i \in N(i \neq S1), \tag{5.9}$$

active and reactive bounds at the slack bus (Eqs. (3.30), (3.31) and (3.32)) and the main feeder bounds

$$S(i,j,h) \leq S_{l.\max}(i,j), \quad i,j \in N(i \neq j), \tag{5.10}$$

and bounds of the curtailment factors

$$0 \leq \beta_{c.pv}(lpv,h) \leq 1. \tag{5.11}$$

5.1.3 A Case study

The formulated optimization problem above is solved for the low-voltage AND shown in Fig. 5.3. The network data is same as considered in the previous chapter. We assume that there is a PVS being connected at each load bus, i.e., at buses 3,4,5,6,7,13,17,20,25,26,27, and 29. Each PVS has a capacity equal to 9 kVA. These assumptions are made based on the study in [96]. The bus number 1 is selected as the slack bus (1.05 fixed voltage amplitude and 0 phase angle), whereas the rest are considered as PQ buses (the upper and lower limits of the voltage amplitude are 1.06 and 0.94, respectively).

Real data of a PVS from a city in Germany is used as PV power penetrations. The profiles of total hourly power for PVSs and demand are shown in Fig. 5.4.

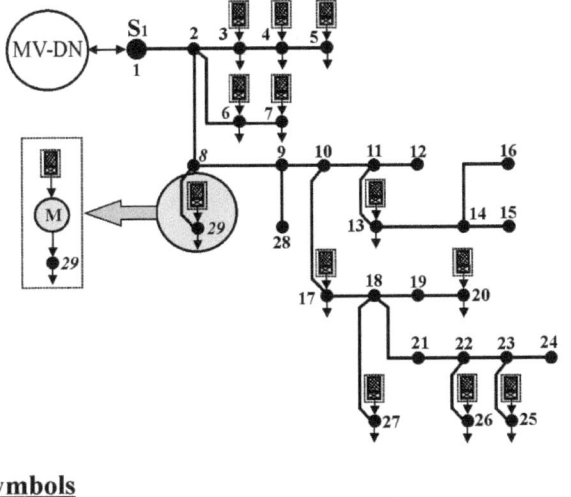

Symbols

↓ Demand   • PQ bus   — Feeder   ●^{S_1} Secondary bus
▓ PVS   ▓ PV unit   ⊠ PCS unit   (M) Meter

Figure 5.3: Low-voltage network for the case study as ADN.

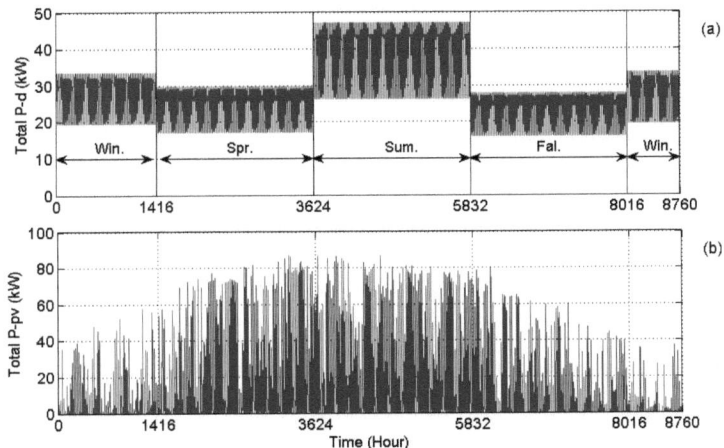

Figure 5.4: Hourly power profiles: (a) Total demand active power. (b) Total PV power generation.

The $C_{pr.art.p}$ is chosen as 0.40$/kWh [41], whereas the on- and off-peak prices $C_{pr.p}$ are assumed to be 0.10$/kWh and 0.05$/kWh [41], respectively. The problem is solved by using GAMS. The optimization problem which is defined in Fig. 5.2 is solved for two modes. In the switch-off mode, the reactive power capability are deactivated, i.e., $Q_{disp.pv} = 0$. In the switch-on mode, $Q_{disp.pv}$ is defined as an optimization variable.

The results in Table 5.1 show that a significant amount of saving can be gained in the total objective by the switching on mode. This gain is obtained mainly from the first term F_{pv}, where a large amount of PV power and energy will be lost when the first mode, i.e., without Q-dispatch, is used. The second term F_{loss} has an unexpected value in the second mode, i.e., with-Q-dispatch. It is commonly recognized that controlling the reactive power dispatch would lead to minimized energy losses, but here our results show that it leads to an increase of energy losses. This interesting point comes from the fact that the increase in the first term F_{pv} is much greater than the decrease in the second term F_{loss}. The total effect leads to the result that a more gain can be achieved. The value of the third term F_{demand} is not changed, because the demand is the same in the first and second mode. Our aim of calculating F_{demand} is for a short-term (one year) analysis.

Figs. 5.5(a)-(b) show the voltage amplitude and voltage angle profiles at bus 17 during typical season's days in a year. It can be clearly seen that the voltage amplitude rises extremely during midday hours. Figs 5.5(c)-(d) show the import and export of the active and reactive power of the network. The negative part during midday hours means that the energy is being exported through the slack bus. In addition, it can be seen that more export can be gained with the with-Q dispatch mode.

Table 5.1: Results of the analysis for the average remuneration tariff (0.40 $/kWh)

Criterion	Without Q-dispatch	With Q-dispatch	Difference
Total objective ($/year)	13100	16839	3739 (**+28.54%**)
F_{pv} ($/year)	34699	38858	4159 (**+11.98%**)
F_{loss} ($/year)	779	**1199**	-420 (**-53.91%**)
F_{demand} ($/year)	20820	20820	0.0 (**0.0%**)
$S_{PCS.max.b}$ (kVA)	37	**0.0**	37 (**100%**)
E_{BSS} (kWh)	207	**0.0**	207 (**100%**)

It is interesting to see in Fig. 5.5(d), that a "huge" amount of reactive power needs to be imported. This amount is needed to cover the reactive power absorption from the PV-PCSs inside the network. In other words, these PV-PCSs work during midday hours of heavily sunny days as an inductive demand and absorb a huge amount of reactive energy, as seen in Fig. 5.6(a), in order to hold the voltages at the buses and keep them within their predefined operation ranges.

Fig. 5.6(b) shows the active power losses during the same selected days. It is clearly seen that a large amount of losses occurs during days in spring. This means that the solution provides a strategy to convert a large amount of PV active energy to energy losses instead of curtailing or spilling it out from the network. In other words, this converted energy is being remunerated by governmental regulations, because it passes through the meters (e.g., at bus 29, as seen in Fig. 5.3) and is regarded as PV energy generation. Another important conclusion from the optimization is that a BSS is not needed to accommodate spilled energy, because there are no curtailments when using the with-Q-dispatch mode. But if reactive power dispatch is not considered, there will be power curtailments and spilled energy, as shown in Figs. 5.7(a)-(b).

$S_{PCS.max.b}$ and E_{BSS} in Table 5.1 show the maximum power curtailed at an hour and maximum energy spilled at a day, respectively. $S_{PCS.max.b}$ and E_{BSS} are used as criteria, for instance in [7], to indicate a BSS size. Now, it is clearly seen that no BSSs are needed to accommodate such spilled energy. It is worth mentioning here that reactive energy to be imported is much cheaper than the PV energy [125].

5.1.4 Conclusions

We formulated a mathematical model for A-R-OPF to analyze low-voltage ADNs with a high penetration of PV-based DG units. The goal is to reveal the impact of controlling and utilizing DG reactive power capabilities on the operations of DNs. Some interesting points have been found and we discussed the results through a case study. It is shown that a huge increase in the total revenue (more than 25%) can be gained if the reactive power capability of PV-based DG units is optimally utilized. Moreover, there is no need in this case study to install BSSs to accommodate spilled PV active energy, because no further spilled energy is present. The aspects and results presented in this section can be used for planning and operating future power networks.

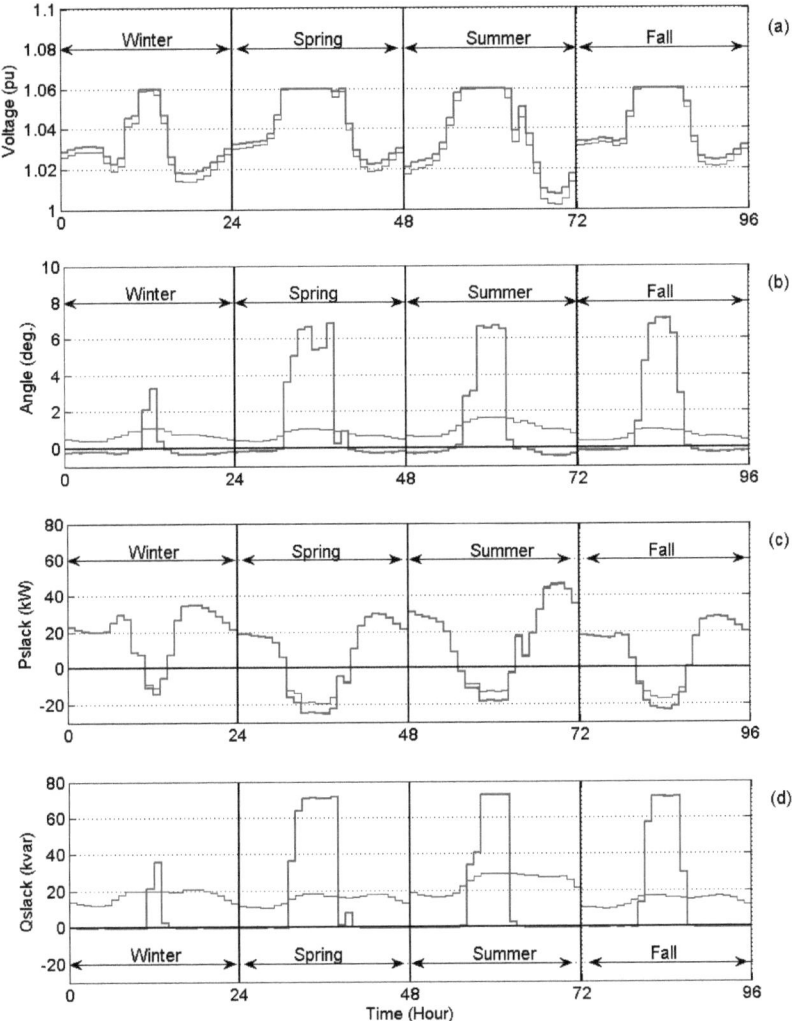

Figure 5.5: Typical four days: (a) Voltage amplitude and (b) voltage angle of bus No. 17. (C) Active and (d) reactive power import/export at slack bus. Note: from (a) to (d) the thin-blue lines stand for the option without Q-dispatch and the bold-red lines stand for the option with Q-dispatch.

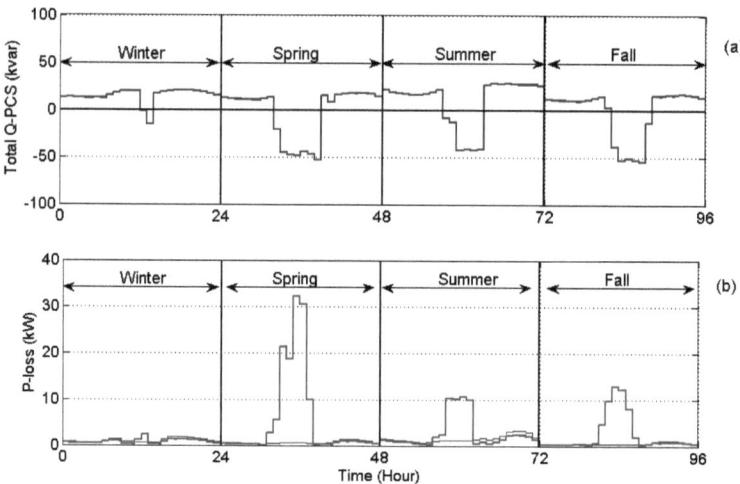

Figure 5.6: (a) Total reactive power dispatch of PV-PCSs. (b) Power losses in typical four days: the (thin-blue) line stands for the option without Q-dispatch and the (bold-red) line for with Q-dispatch.

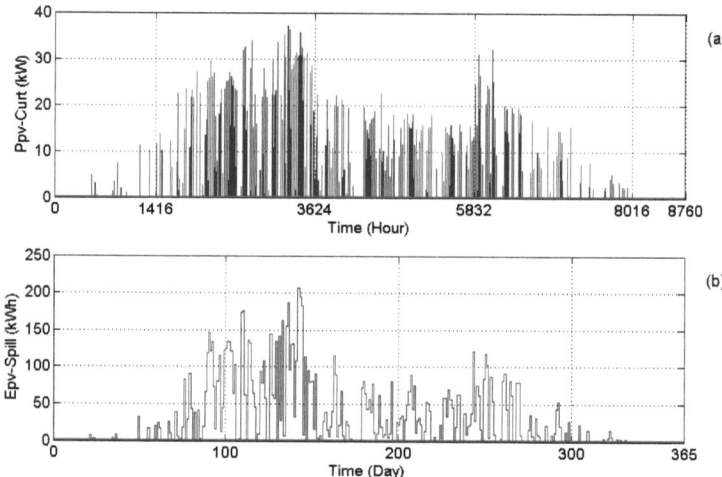

Figure 5.7: (a) Total PV power curtailments. (b) Total PV spilled energy. Here, the mode of without Q-dispatch (blue) and with Q-dispatch (red).

5.2 A-R-OPF for Medium-Voltage ADNs

5.2.1 Modeling of Network Demand, Generation and Energy Prices

The load model is taken as explained in Section 3.2.1, whereas the wind power penetration as given in Section 3.3.1. The price model is considered to be a two-tariff price model as depicted in Fig. 5.8. Here, the energy prices are low during night hours (T_1 and T_3) and high during day hours (T_2). Note that reactive energy prices are neglected in this case study.

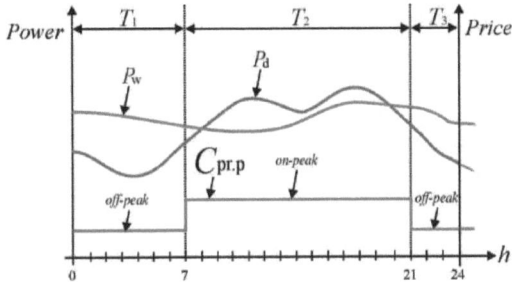

Figure 5.8: Daily wind-demand power profiles and active energy price model for the medium-voltage ADN with wind stations and BSSs. Here, $C_{pr.p}$ stands for a two-tariff price model of active energy.

5.2.2 A-R-OPF with Wind-Battery Stations

In this section, we present a mathematical model and formulate an optimization problem for the medium-voltage ADN with embedded wind generation and battery storage. A power system to be optimized can be described with Fig. 5.9. Here, the wind power and the demand profiles are given as inputs to the system. The control variables (active power charge and discharge, reactive power dispatch and wind power curtailment factor) will be optimized to achieve an optimal operation for total revenue maximization and total losses minimization.

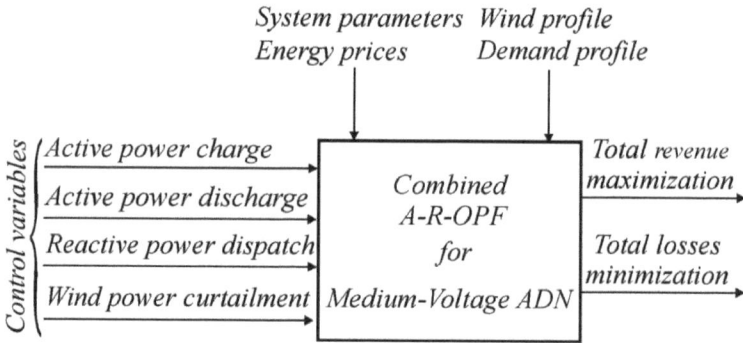

Figure 5.9: Input-output scheme for the combined A-R-OPF for the medium-voltage ADN.

For comparison, we use the same control variables in the OPF model as in the A-R-OPF model, but in the OPF model the reactive power dispatch is set to zero (i.e., $Q_{disp.b}(lb,h) = 0$), where lb is the set of BSSs. As shown in Section 3.3.1, the aim of introducing a curtailment factor for wind power at each wind park is to ensure a feasible solution, i.e., to spill a part of wind energy out, when the capacity of the installed BSSs are not enough or other system constraints will be violated. It means, if no wind power will be curtailed, then $\beta_{c.w}(lw,h) = 1$, otherwise $\beta_{c.w}(lw,h) < 1$, where lw is the set of wind parks. Another way of optimally utilizing the potential of the system is to consider a longer time horizon, if the demand and wind power profile in the time horizon can be forecasted. It is noted in most previous studies on OPF that the time horizon for planning DN operations has been always 24 hours. However, when embedded BSSs are available, a multi-day strategy can be used for the optimization to exploit their storage capacity for a higher benefit. The mathematical model of the combined A-R-OPF problem is given as follows.

Objective function:

The objective function is defined as follows

$$\max_{P_{ch}, P_{dis}, Q_{disp.b}, \beta_{c.w}} F_{w.b} - F_{loss}, \quad (5.12)$$

where

$$F_{w.b} = \sum_{h=1}^{T_{final}} C_{pr.p}(h) \sum_{\substack{i=1 \\ i \in lw, lb}}^{N} (P_w(i,h)\beta_{c.w}(i,h) + P_{dis}(i,h) - P_{ch}(i,h)) \quad (5.13)$$

and

$$F_{loss} = \frac{1}{2}\sum_{h=1}^{T_{final}} C_{pr.p}(h) \sum_{i=1}^{N}\sum_{j=1}^{N} G(i,j)\Big(\left(V_e(i,h)\right)^2 \\ + \left(V_f(i,h)\right)^2 + \left(V_e(j,h)\right)^2 + \left(V_f(j,h)\right)^2 \\ - 2\left(V_e(i,h)\,V_e(j,h) + V_f(i,h)\,V_f(j,h)\right) \Big). \quad (5.14)$$

Eq. (5.13) describes the total revenue from the wind energy and from the BSSs, whereas (5.14) represents the total cost due to the energy losses. In order to avoid confusion due to the complex relationship between relevant parties in the liberalized electricity market [119][7], the following simple market strategy is considered.

- The integration of BSSs in an ADN with a high penetration of wind-based DG units is economically feasible.
- The wind-based DG units, BSSs embedded in an ADN, reverse active energy to the TN and active energy losses are paid for or charged by the same price model, i.e., the two-tariff price model.
- Both active and reactive reverse power flow to the TN is allowed without any rejection.

Equality constraints:

The active power flow equations at the buses are given as follows:

$$V_e(i,h)\sum_{\substack{j=1\\j\in i}}^{N}(G(i,j)V_e(j,h)-B(i,j)V_f(j,h))$$
$$+V_f(i,h)\sum_{\substack{j=1\\j\in i}}^{N}(G(i,j)V_f(j,h)+B(i,j)V_e(j,h))+P(i,h)=0, \quad i\in N, \tag{5.15}$$

where the active power injection at bus i during hour h is calculated by

$$P(i,h)=P_d(i,h)-P_w(i,h)\beta_{c.w}(i,h)-P_{S1}(h)-P_{dis}(i,h)+P_{ch}(i,h). \tag{5.16}$$

The reactive power flow equation is given by

$$V_f(i,h)\sum_{\substack{j=1\\j\in i}}^{N}(G(i,j)V_e(j,h)-B(i,j)V_f(j,h))$$
$$-V_e(i,h)\sum_{\substack{j=1\\j\in i}}^{N}(G(i,j)V_f(j,h)+B(i,j)V_e(j,h))+Q(i,h)=0, \quad i\in N \tag{5.17}$$

where the reactive power injection at bus i during hour h is calculated by

$$Q(i,h)=Q_d(i,h)-Q_{\text{disp.b}}(i,h)-Q_{S1}(h) \qquad i\in N. \tag{5.18}$$

In addition, energy balance equations for BSSs are also included (see Eqs. (3.20) and (3.21)).

Inequality constraints:

The inequality constraints consist of the satisfaction of voltage bounds

$$V_{\min}(i)\le V(i,h)\le V_{\max}(i) \qquad i\in N(i\ne \text{S1}), \tag{5.19}$$

active and reactive bounds at the slack bus (see Eqs. (3.30), (3.31) and (3.32)) and the main feeder bounds

$$S(i,j,h) \leq S_{l,\max}(i,j), \qquad i,j \in N(i \neq j) \qquad (5.20)$$

and the bounds of the curtailment factors

$$0 \leq \beta_{c,w}(lw,h) \leq 1. \qquad (5.21)$$

The restrictions of control variables (see Eqs. (3.15)-(3.19)) and capacity limits of the BSSs inequality constraints (see Eq. (3.22)) should also be included into inequality constraints.

The A-R-OPF formulated above leads to a large-scale NLP problem consisting of three control variables for each BSS and one for each wind park embedded in the considered ADN in the time horizon. The state variables included in the problem are the energy storage level of each BSS, real and imaginary components of the complex voltage, and active as well as reactive power at the slack bus in the time horizon, respectively. We solve this problem with GAMS. The computation is carried out on a desktop with Intel Core i7 CPU 3.37 GHz and 3.25 GB RAM.

It should be noted here that the choice of the initial values in the A-R-OPF method has an impact on both the feasibility and computational effort. The initial values, denoted by (0), chosen in this case study for all computations are the usual flat start Pch(0) = Pdis(0) = Qdisp(0) = Vf(0) =PS1(0) = QS1(0) = E(0) = 0 and βc.w(0) =Ve(0) = 1. Many different initial values have been tested. It converges to the same results, but the CPU time is different. Only when the initial values are very far from the flat start a convergence problem may occur. From another perspective, the result from the combined A-R-OPF given in Fig. 5.9 is given to the DSI-1 (see in Appendix D) to verify the results produced by the NLP solver and evaluate the Jacobian matrix for well- and ill-conditioned power systems [40].

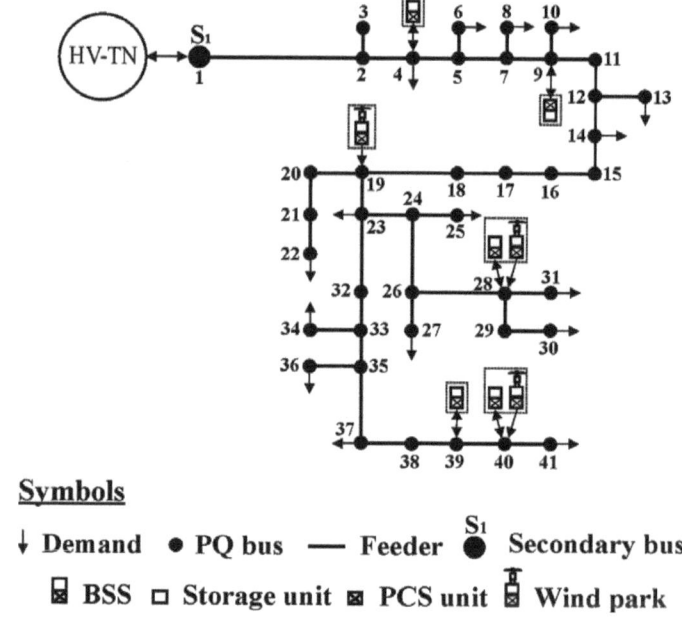

Figure 5.10: Medium-voltage network for the case study as ADN.

5.2.3 A Case study

In this section, the proposed method is applied to the medium-voltage ADN as seen in Fig. 5.10. This ADN is studied in [7], but the reactive power of BSSs was not considered. Our aim is to demonstrate the potential when the reactive power dispatch is also considered in the OPF problem. The data of the demand, wind turbines, PCSs capabilities and BSSs capacities are given in Appendix C: Table C.2, Table. C.3 and Table C.4. The ADN has three embedded wind parks connected to buses 19, 28, and 40, respectively. The wind power speed is assumed to be the same at the wind parks. Five BSSs are connected to buses (4, 9, 28, 39 and 40) which were also considered in [7]. Bus 1 is considered to be the slack bus (1.0 fixed voltage amplitude and 0 phase angle), whereas other buses are considered to be PQ buses. The on- and off-peak prices are assumed to be 100$/MWh and 50$/MWh, respectively.

The total wind power and demand scenarios in five days considered for the optimization are shown in Figs. 5.11(a) and 5.12(a). The wind profile is chosen from five days in spring of a German city. To make a clear comparison, we consider two cases, one for A-R-OPF (i.e., with reactive power dispatch) and one for commonly applied OPF (i.e., without reactive power dispatch). In addition, we solve the optimization problem with two different optimization horizons (procedure A and procedure B). In procedure A, the problem has a daily horizon (T_{final} = 24h) and is solved five times for the five individual days. In this procedure, the final storage level is required to be equal to the initial storage level for each day. In procedure B, the time horizon is defined as five days (T_{final} =120h) and the problem is solved only once, and the final storage level should be equal to the initial value of the time horizon.

The optimization results are shown in Table 5.1 to Table 5.4 and in Figs. 5.11 and 5.12. It can be seen in Table 5.1 that the differences in the total revenue between procedure A and B and in the case with and without reactive power dispatch are not considerable, since the total revenue from wind energy and BSSs dominates. However, if we examine the total losses, as shown in Table 5.2, a significant reduction can be achieved by the A-R-OPF strategy. It means that this loss reduction will dominate in days without wind power.

A huge reduction in the total reactive energy imported from the TN can be gained, as shown in Table 5.3.

Table 5.1: Total revenue for procedure A & B in two cases

	A-R-OPF	OPF	Difference
A($)	35082	34909	173(0.49%)
B($)	35473	35302	171(0.48%)
Difference($)	391(1.11%)	393(1.12%)	---

Table 5.2: Total costs of losses for procedure A & B in two cases

	A-R-OPF	OPF	Difference
A($)	1154	1314	160(12.17%)
B($)	1123	1290	167(12.94%)
Difference($)	31(2.68%)	24(1.82%)	---

Table 5.3: Slack bus reactive energy for procedure A & B in two cases

	A-R-OPF	OPF	Difference
A(Mvarh)	28.163	376.528	348.365 (92.52%)
B(Mvarh)	24.593	374.091	349.498 (93.42%)
Difference (Mvarh)	3.570(12.67%)	2.437(0.64%)	---

Table 5.4: Number of variables (N.V.) and computation time (CPU) for procedure A & B in two cases

	A-R-OPF		OPF	
	N.V.	CPU (sec.)	N.V.	CPU (sec.)
A	2520	162	2400	56
B	12600	680	12000	227

In order to analyze the impact of the injected reactive power from the PCSs on the reactive energy imported/exported from/to the TN, a comparison between OPF ($Q_{disp.b}(lb,h) = 0$) and A-R-OPF is made. Since no reactive power sources are used in the OPF, this leads to the high difference in the values between the A-R-OPF and OPF in Table 5.3.

The slack bus reactive energy is calculated by $\Sigma_h\, Q_{S1}(h)$, where Q_{S1} is high and positive (i.e., imported see Fig. 5.12(c) dashed line for procedure B) in the OPF (due to $Q_{disp.b}(lb,h) = 0$), while it is small and positive or negative (i.e., imported or exported see Fig. 5.12(c) bold and thin lines) in the A-R-OPF.

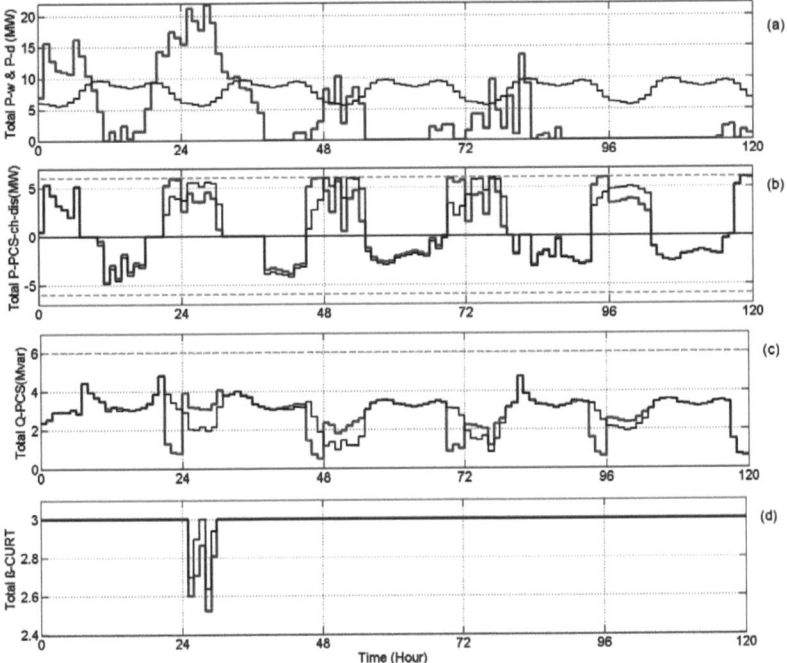

Figure 5.11: Operation strategies of A-R-OPF: (a) total wind power generation (bold-blue) and total demand power (thin-black). (b) Total active power charge/discharge. (c) Total reactive power dispatch. (d) Total curtailment factor. Note: from (b) to (d) the lines (bold-blue) for procedure A and (thin-black) for procedure B.

Therefore, a significant reduction can be seen in Table 5.3, which is in agreement with the recent result in [80] where control schemes for reactive power in DNs were studied but without considering BSSs. From the above results it can be concluded that significant installation costs [62] of devices for compensating the needed reactive power can be saved using the proposed approach compared with the OPF strategy. This saving can be used to cover a considerable part of installation costs for BSSs, i.e., this will increase the value of integrating BSSs in DNs.

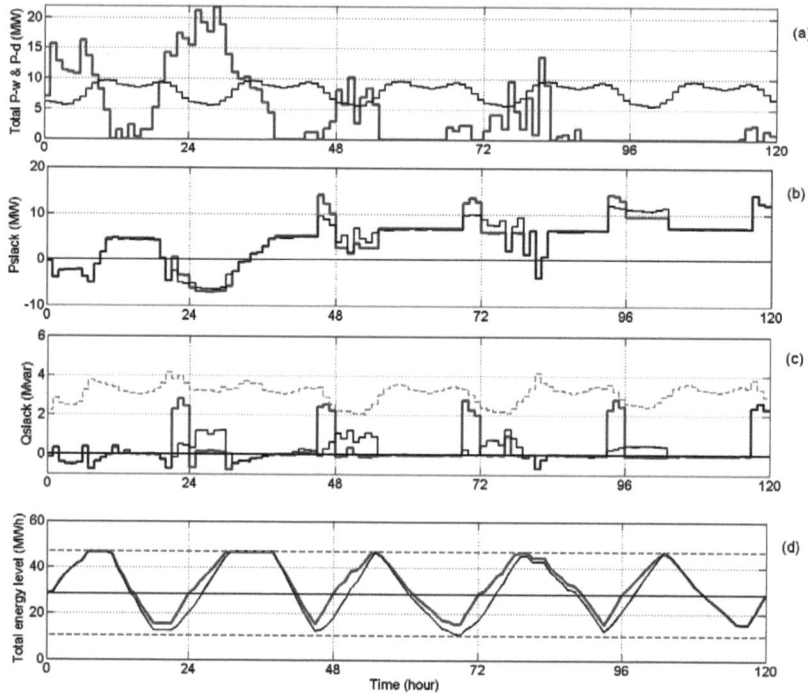

Figure 5.12: Operation strategies of A-R-OPF: (a) Total wind power generation (bold-blue) and total demand power (thin-black). (b) Slack active power. (c) Slack bus reactive power (dashed-red line for procedure B and OPF). (d) Total energy level. Note: from (b) to (d) the lines (bold-blue) for procedure A and (thin-black) for procedure B.

It is also interesting to see the differences of the results from procedure A and B shown in Table 5.1 to 5.3. It is indicated that benefits can always be obtained if a five-day optimization horizon is chosen, instead of the daily optimization for five individual days. As a result, a long time horizon should be used to utilize the storage potential of BSSs, when forecasting data are available. In Table 5.4, the CPU time take to run the optimization program is given. In Figs. 5.11(b)-(d) and 5.12(b)-(d) the optimal profiles of A-R-OPF from procedure A (bold) and B (thin) are compared. The optimal control strategies for the total active power charge/discharge of the five PCSs are shown in Fig. 5.11(b), where the positive value means charge and negative value discharge. In the first four windy days the active power charge/discharge corresponds to the available wind power, while in the fifth day BSSs should also be charged/discharged due to the two-tariff price model.

In Fig. 5.11(c) it can be seen that a large amount of reactive power is optimally utilized. The total curtailment factor for the three wind parks is shown in Fig. 5.11(d). It is shown that in almost all of the time the wind power will not be curtailed. Only in the second day, in which a large wind power penetration is available, a small part of it will be curtailed due to the system constraints.

Fig. 5.12(b) shows the power imported from and exported to the TN. The impact of the two-tariff prices can be clearly seen from the profiles in the fifth day. From Fig. 5.12(c) it is shown that only a small amount of reactive energy will be needed from the TN because of the reactive energy compensated by the BSSs, as shown in Fig. 5.11(c). From Fig. 5.12(c) there is an amount of negative reactive power. This means that such reactive power can be exported back to the TN when the reactive power capability of the B-PCSs is more than enough to cover the total reactive demand as well as reactive power losses inside the network. The total energy level stored in the five storage units is shown in Fig. 5.12(d). It can be seen that the storage capacity will be fully exploited when necessary. The advantages of using a long time horizon ($T_{final} = 120h$) compared with a short time horizon ($T_{final} = 24h$) can be seen from the thin and bold lines in Figs. 5.11(b)-(d) and 5.12(b)-(d).

It is indicated from Figs. 5.11(b)-(c) and 5.12(b)-(c) that the power profiles are smoothed by the 5-day operation strategy, while these are more fluctuating by the daily operation strategy.

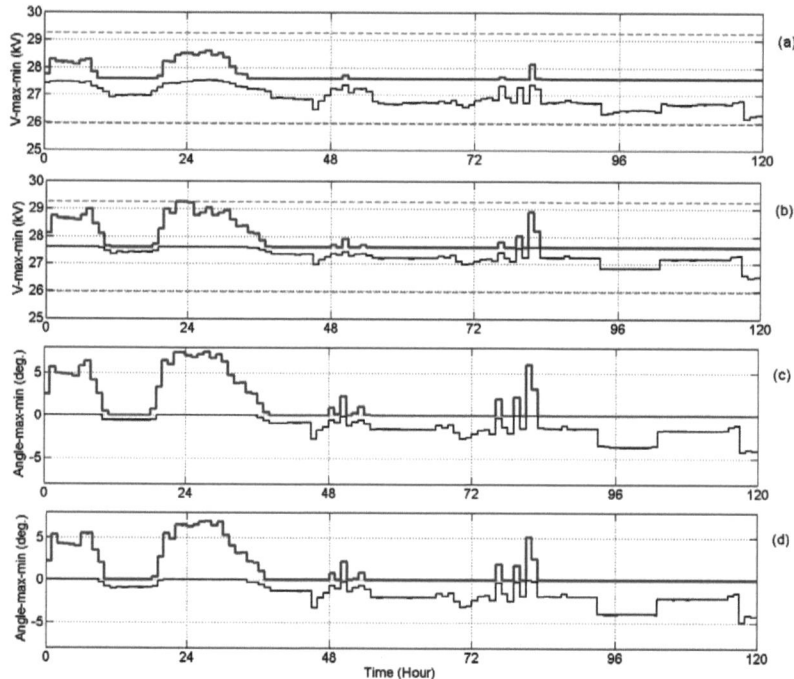

Figure 5.13: Operation strategies of A-R-OPF and OPF for procedure B: (a) The maximum voltage amplitude of buses (bold-blue) and minimum (thin-black) for OPF and (b) for A-R-OPF. (c) The maximum voltage angle of buses (bold-blue) and minimum (thin-black) for OPF and (d) for A-R-OPF.

From Fig. 5.11(d) the curtailed wind power is lowered by the long horizon strategy in comparison to that of the short horizon strategy. From Fig. 5.12(d) it can be seen that the storage level by the long horizon strategy is always lower than that from the short horizon strategy. It means that using a longer time horizon will lead to a higher degree of employing the storage capacity.

Fig. 5.13(a) shows the maximum and minimum voltage amplitude among all buses in OPF without reactive power dispatch ($Q_{disp,b} = 0$), while Fig. 5.13(b) with reactive power dispatch. It can be seen that using the A-

R-OPF both of the maximum and minimum voltage values increase and the profiles are more fluctuating in comparison to those from OPF without reactive power dispatch. The maximum and minimum voltage angle among all buses for ($Q_{disp.b} = 0$) is depicted in Fig. 5.13(c), while Fig. 5.13(d) is with reactive power dispatch. Comparing Fig. 5.13(c) and Fig. 5.13(d), it is shown that the reactive power dispatch has almost no impact on the maximum and minimum voltage angle of the buses, which is in agreement with the result of [97].

The optimal control profiles obtained from the NLP solver and the other input parameters are provided to the DSI-1 as inputs, while the initial state variables are chosen to be flat initial values as provided to the NLP solver. Results obtained from the DSI-1 are identical to those from the NLP solver.

As seen from Chapter 4, the power flow Jacobian matrix of the case has a high condition number [112] bounded by 2×10^4 in the considered time horizon. However, in the computations of procedures A and B for the given scenarios and with the usual flat start no convergence problems have been seen in solving the case study using the mentioned NLP solver. Moreover, not more than five iterations are needed for the DSI-1 to converge to the results produced by the NLP solver when the usual flat start is used.

5.2.4 Conclusions

Many issues have impacts on the operations of DNs. But in previous studies these issues have been considered separately. In this section, we have proposed a new, combined problem formulation for A-R-OPF of ADNs with embedded wind power generation and battery storage. Both active and reactive power dispatch strategies will be optimized for maximizing the total revenue and simultaneously minimizing the total costs of energy losses. The proposed A-R-OPF is based on a two-tariff price model and one charge/discharge cycle in each day. In addition, we have considered and compared two different optimization horizons, i.e., one-day strategy and multi-day strategy.

A real DN with three embedded wind parks and five BSSs has been used for evaluating the effectiveness of the proposed approach. The results show a reduction of 12% of the energy losses and 90% of reactive energy needed to be imported from the TN can be achieved by the A-R-OPF operation strategy. Moreover, a long optimization horizon leads to benefits not only economically but also in the sense of a smoothing operation, i.e., reducing the fluctuations during the operation.

6 Flexible Optimal Operation of Battery Storage Systems for Energy Supply Networks

From Chapter 5, we conclude that to prolong the life of BSSs only one fixed charge/discharge cycle every day can be considered. However, due to the fact that the profiles of renewable energy generation, demand and prices vary from day to day a fixed operation of BSSs cannot be optimal.

In this chapter, a flexible battery management system is proposed to adapt to such variations. Using this system a considerably higher revenue can be achieved. This is accomplished by optimizing the lengths (hours) of charge and discharge periods of BSSs for each day, leading to a complex MINLP problem. An iterative two-stage framework is proposed to address this problem. In the upper stage, the integer variables (i.e., hours of charge and discharge periods) are optimized and delivered to the lower stage. In the lower stage the A-R-OPF problem is solved by a NLP solver and the resulting objective function value is brought to the upper stage for the next iteration.

6.1 Problem Description

For a clear comparison with the results in Chapter 5 the same medium-voltage ADN (see Fig. 5.10) is considered here. It is necessary to develop an optimal strategy which will lead to a maximum revenue and a feasible operation. Note that the major difficulty in this problem comes from: 1) the dynamic behavior of demand, wind power generation and energy prices and 2) the operational constraints of BSSs. Furthermore, market strategies are also to be taken into consideration. These difficulties are explained in detail in the following sections.

6.1.1 Varying Demand, Generation and Energy Prices Profiles

As shown in Fig. 6.1, demand and wind power profiles are different from day to day. For clarifying this fact two typical demand profiles in different season's days, denoted by $P_d^{(A)}$ (winter) and $P_d^{(B)}$ (spring), are shown in Fig. 6.1. In addition, two wind power profiles (represented by day-ahead forecasted scenarios) are also depicted in Fig. 6.1 and denoted by $P_w^{(A)}$ and $P_w^{(B)}$.

In contrast, time-of-use (TOU) pricing is usually used by utilities for charging different rates throughout the day [37], [124]. It means that active energy prices are low during low demand and high during high demand. This is depicted in Fig. 6.2, where $C_{pr.p}(h)$ is the active energy price during hour h. In this chapter, three different price models for active and reactive energy prices, as shown in Fig. 6.2, are used to analyze their effect on the operation.

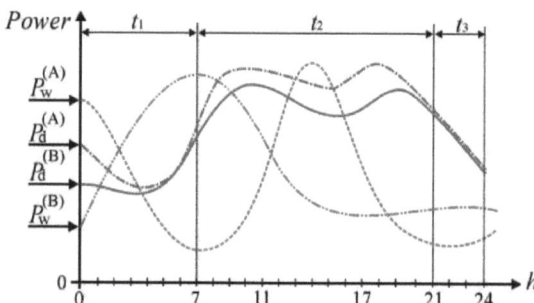

Figure 6.1: Daily wind/demand power profiles, where $^{(A)}$ and $^{(B)}$ stand for different wind/demand power profiles at different days.

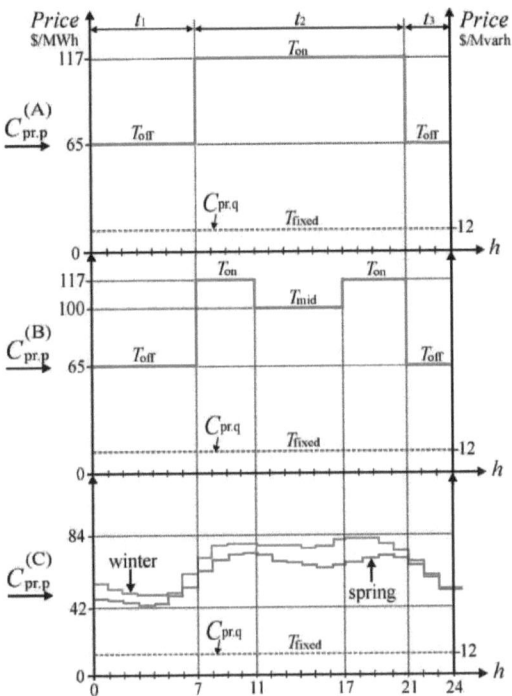

Figure 6.2: Daily active and reactive energy price models for the MV level with wind stations and BSSs. Here, (A), (B), and (C) stand for a two-/three-/24-hour-tariff price model of active energy, respectively, while the fixed line (dashed) stands for a fixed-tariff price model of reactive energy.

Here, $C_{pr.p}^{(A)}$, $C_{pr.p}^{(B)}$, and $C_{pr.p}^{(C)}$ denote the two-/three-/and 24-hour-tariff price model, respectively. T_{off}, T_{mid}, and T_{on} stand for the durations of low, medium and high prices for the first two price models, respectively. Different hourly prices are assumed to follow the demand in winter and spring for the third price model.

In addition, reactive energy prices are applicable in certain countries based on the measured reactive energy (Mvarh) [125] (e.g., in Germany), or based on the costs of providing reactive power, including additional costs due to energy losses incurred by running at a non-unity power factor and costs of running the generation units as synchronous condensers if requested by the independent electricity system operator (IESO) [16] (e.g., in Ontario). It is suggested in [125] [16] that reactive energy prices can be assumed in the range of 3-13 $/Mvarh. In this chapter, a fixed reactive energy price T_{fixed}, as seen in Fig. 6.2, is considered for comparison purposes.

All of these profiles are important to the operation of the distribution network. They are time dependent and different from day to day. Therefore, operation strategies for BSSs should be flexible and adaptive to the variations of these profiles so as to always ensure the system in an optimal and reliable performance.

It should be noted that the above mentioned wind power and demand profiles are assumed to be forecasted in the time frame of optimization as will be shown later. The inaccuracies in these forecasts are not considered in this study.

6.1.2 Operational Constraints of BSSs

It was shown in [89] [87] that the lifetime of a BSS depends on a fixed number of charge/discharge cycles and DoD. This can be represented by a replacement period r in years as follows [87]

$$r = \frac{p}{n \times D} \quad (6.1)$$

where p is the total number of charge/discharge cycles in the lifetime, D is the annual operation days, and n is the number of charge/discharge cycles per day. Therefore, to prolong the lifetime for BSSs only one fixed cycle charge/discharge per day is typically chosen for optimal planning [7] and operation (as shown in Chapter 5).

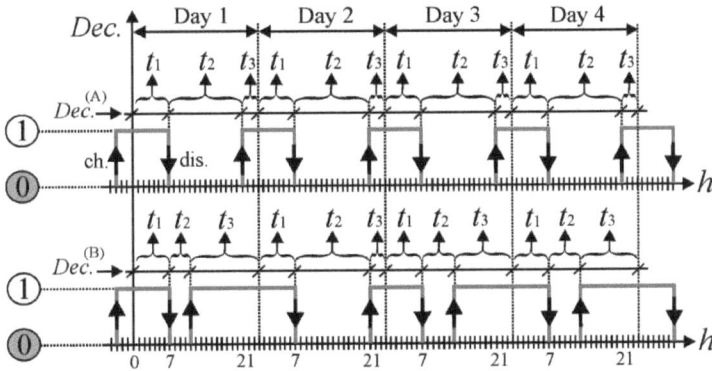

Figure 6.3: Illustration for one charge/discharge cycle every day with two different decisions. Here, (A) and (B) stand for fixed/flexible operations, respectively.

However, when renewable energies as well as other parameters vary from day to day one flexible cycle charge/discharge per day should be considered. Briefly, the cycle of charge is determined by two integer variables representing the time periods (hours) of charge (t_1 and t_3). The cycle of discharge is defined by one integer variable representing the hours of discharge (t_2), as depicted in Fig. 6.3, where the notation 1 means charging period and 0 discharging period. Two operation strategies are shown: the upper part shows a fixed strategy, i.e., stringent decisions, and the lower part a flexible strategy, i.e., free decisions. In this chapter, the lengths of the charge and discharge cycles will be optimized based on the day-to-day profiles discussed above. The three integer variables in a cycle are constrained by

$$t_1 + t_2 + t_3 = t_{max}, \qquad (6.2)$$
$$t_{min} \le t_1 \le t_{max}, \, t_{min} \le t_2 \le t_{max}, \, t_{min} \le t_3 \le t_{max}, \qquad (6.3)$$

where t_{min} and t_{max} are the minimum and maximum bounds on the variables, respectively. Since we consider a daily operation of BSSs, there must be

$t_{min} = 0$ and $t_{max} = 24$. It is noted that two procedures A and B are used in this chapter as in Chapter 5. In procedure A, the problem has a daily horizon ($T_{final} = 24h$) and is solved four times for the four individual days. In procedure B, the time horizon is defined as four days ($T_{final} = 96h$) and the problem is solved only once. In both procedures, the final storage level should be equal to the initial in the time horizon.

6.1.3 Market Strategies

The same simple market strategy defined in Section 5.2.2 is considered in this chapter to make a clear comparison. In addition, three different energy price models, as shown in Fig. 6.2, are used and corresponding results compared. Here, the ADN with distributed wind parks and BSSs is considered to be operated by a DSO who is responsible for operating the system with a high quality. The DSO tends to maximize the benefits from the system operation and meanwhile to minimize the total costs. For a clear analysis following assumptions are taken in the formulated optimization problem.

- The integration of a BSS in a DN with high penetration of DG units is economically feasible.
- The DG units, BSSs embedded in a DN, reverse active energy to the TN and active energy losses are paid for or charged by the same price model, i.e., the two-/three-/24-hour-tariff price model.
- The reactive energy import/export from/to the TN is paid for or charged by the same price model, i.e., the fixed reactive energy price.
- Both active and reactive reverse power flow to the TN is allowed without any rejection.

6.2 Problem Formulation and Solution Framework

6.2.1 Problem Formulation

The optimization problem, as shown in Fig. 6.4, has 3 integer variables in addition to the continuous control variables (i.e., three control variables for each BSS and one for each wind park, as given in Section 5.2.2). The time step is 1 hour for each continuous variable.

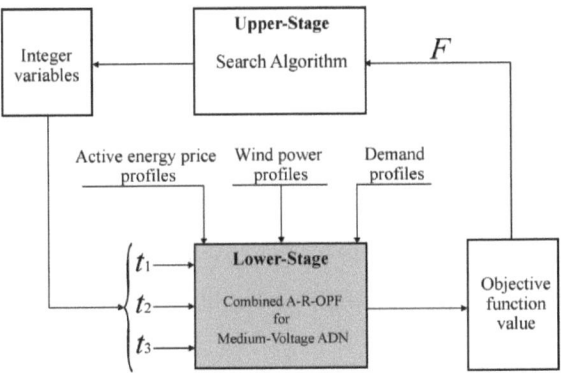

Figure 6.4: Input-output scheme for the combined A-R-OPF for the medium-voltage ADN with a search algorithm.

A general formulation of an A-R-OPF problem can be expressed as follows

$$\max_{x,u,t} \quad F(x,u,t), \tag{6.4}$$

$$\text{s.t.} \quad g(x,u,t) = 0, \tag{6.5}$$

$$x_{min} \leq x \leq x_{max}, \tag{6.6}$$

$$u_{min} \leq u \leq u_{max}, \tag{6.7}$$

where the objective function F (see Eq. (5.12)) to be maximized is the total revenue from wind power and BSSs (see Eq. (5.13)) minus the total cost of energy losses (see Eq. (5.14)), x is the vector of state variables (real and imaginary component of complex voltage at PQ buses, active and reactive power injected at slack bus and energy level in BSSs), u is the vector of continuous control variables including active power charge/discharge of BSSs, reactive power dispatch of BSSs and curtailment factors of wind power at wind parks, t is the vector of the integer control variables, i.e., the number of charge/discharge hours per day.

In Eq. (6.5), g represents equality constraints including active and reactive power flow equations (see Eqs. (5.15), (5.16), (5.17), and (5.18)). In addition, energy balance equations for BSSs are also included (see Eqs. (3.20) and (3.21)). The inequality constraints include voltage bounds (see Eq. (5.19)), active and reactive bounds at the slack bus (see Eqs. (3.30), (3.31) and (3.32)), the main feeder bounds (see Eq. (5.20)), and the bounds of the curtailment factors (see Eq. (5.21)). The restrictions of control variables (see Eqs. (3.15)-(3.19)) and capacity limits of the BSSs (see Eq. (3.22)) should also be included into inequality constraints.

6.2.2 A Two-Stage Solution Framework

To solve the MINLP problem formulated above an iterative two-stage framework scheme is proposed, as shown in Fig. 6.4. The whole optimization problem is decomposed into two sub-problems. In each iteration, the upper stage solves the following problem

$$\max_{t} F(x(t), u(t), t) \qquad (6.8)$$

subject to Eqs. (6.2) and (6.3), where only the integer variables t are search variables. With t values delivered from the upper stage, the lower stage solves the following NLP problem

$$\max_{x,u} F(x, u), \qquad (6.9)$$
$$\text{s.t.} \quad g(x, u) = 0, \qquad (6.10)$$

and Eq. (6.6) and Eq. (6.7) as inequality constraints, where only continuous variables are present. The solution of the lower stage provides the objective function value for the upper stage where an update of t will be made for the next iteration. This procedure will converge when a number of iterations is reached as described below.

The influence of energy prices, wind power generation and demand profiles is dealt with in the lower stage. Thus the lower stage solves the A-R-OPF problem with given charge/discharge hours provided from the upper stage. A unique feature of the lower stage is that the operational constraints will be ensured for any charge/discharge lengths, due to the introduction of curtailment factors. Therefore the lower stage can be considered as a black-box solver for the upper stage.

Here, the upper stage is implemented in MATLAB while the lower stage in GAMS. The framework shown in Fig. 6.4 is realized by using GDXMRW for interfacing GAMS and MATLAB. In this chapter, the computation is carried out on a desktop with Intel XEON X5690. 3.47 GHz (6-core) 32.00 GB RAM.

6.2.3 A Search Method for the Upper Stage Problem

Here, we focus on developing optimal and flexible strategies for operating BSSs. Such strategies can handle not only the continuous decision variables but also integer variables.

Developed by Holland [56] and Goldberg [49], GA has been successfully applied in solving many optimization problems in power systems, especially when both integer and continuous variables are present. In [64], GA was applied to solve unit commitment problems. Additional schemes like intelligent and problem-oriented permutation mechanisms were added to improve the GA search [85].

Chap. 6: Flexible Optimal Operation of Battery Storage Systems for Energy Supply Networks

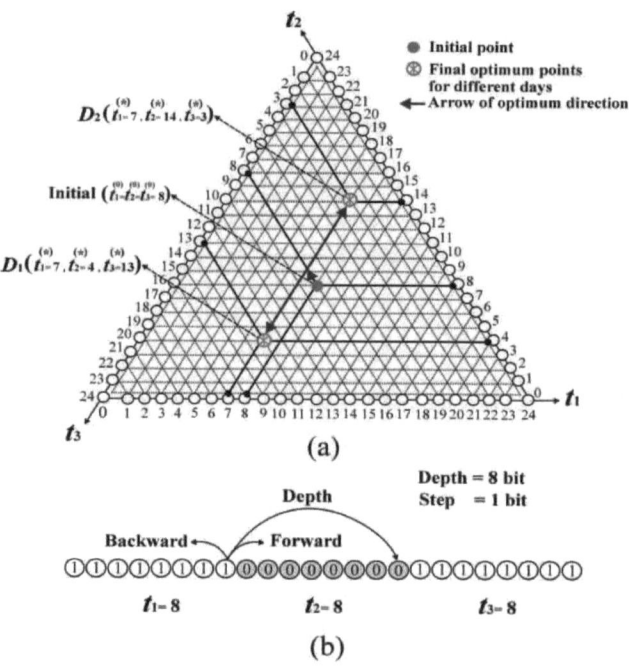

Figure 6.5: (a) Illustration of the search space. (b) The string-structure.

The authors in [64] and [85] presented an enhanced GA [9] for the solution of OPF with both continuous and discrete control variables. Kennedy and Eberhart [66] presented a method for optimization of continuous nonlinear functions using particle swarm optimization. They extended their method to handle binary variables [67]. Such a method is applied in reactive power and voltage control formulated as a MINLP problem [122]. Other search methods such as machine learning [91] and motion estimation [2], [99], were also used in solving optimization problems with integer variables.

Based on the problem formulation in this chapter and due to the constraints of the three integer variables described by Eq. (6.2) and Eq. (6.3), the search space can be illustrated with Fig. 6.5(a), where each point represents a possible combination of the variables. Thus the total number of combinations is 325. Note that formulating the problem with more than one cycle per day or with smaller time steps will increase the complexity of the problem by increasing the search scales. Since an exhaustive search or enumeration will cause much computation time, we use a more efficient search method described as follows.

Goldberg pointed out in [49] that a G-bit improvement is a canonical method of local search which could be hybridized with GAs. This method can be summarized by a bit-by-bit sweeping process in a given string, see Fig. 6.5(b).

In contrast, Adby and Dempster [2] presented a simple technique in a two-variable optimization case (see an application of this method in [99]), in which only one variable is adjusted at a time with the other fixed. Thus it is called one-at-a-time search (OTS). Since there are three variables instead of two in this case and they should satisfy Eq. (6.2) and Eq. (6.3), we modify OTS and use it in the upper stage, including the following steps:

1) Choose a feasible initial string (including 24-bit, either 0 or 1, where 1 stands for charge and 0 for discharge). For example, $t_1^{(0)} = t_2^{(0)} = t_3^{(0)} = 8$ which is depicted in Fig. 6.5(a) and 6.5(b).

2) Provide this initial string to the lower stage to evaluate the objective function value. Then record this fitness in a register.

3) Sweep t_1 bit by bit (the search step is taken to be 1 bit which represents the smallest step) in the forward as well as backward direction, as shown in Fig. 6.5(b), with a specific depth (8 bit which represents the maximum depth when $t_1^{(0)} = t_2^{(0)} = t_3^{(0)} = 8$). Note that t_3 is still fixed at this sweep process, while t_2 is being changed depending on the change in t_1. For example: $t_1 = 10$, $t_2 = 6$, and $t_3 = 8$ after two sweeps in the

forward direction. Note that after each sweep the produced string is evaluated in Step 2.

4) Sort the fitness of the successive evaluated strings in ascending order and retain the best of them (if two or more strings have the same fitness, the algorithm preserves the original ordering of the fitnesses). This gives the best position of t_1 and t_2 related to t_3.

5) Fix t_1 at its best position and begin to sweep t_3 and evaluate the produced strings in a similar way as in Steps 3 and 4. The total number of evaluations needed to find a converged solution is 33, given the parameters (initial point, depth and step) as shown in Fig. 6.5.

6) The best string found and its fitness represents the optimal operations for a specific day.

It should be noted that any sweep which violates the constraints Eq. (6.2) and Eq. (6.3) will be refused. In the integer search method the optimization horizon is set to one day (24 hour) which is then repeated. Since the modified OTS method is a local search scheme, a large number of scenarios were tested in the case study (see Section 6.3) to show the impact of initialization, depth and step of the search. From the computed scenarios following can be observed:

- Local hills can be obtained when a bad initialization is used, e.g., starting from corners of the search space.
- If the search step is more than one bit maximum hills could be prevented.
- The parameters (initial point, depth and step) are logically related. It means if the initialization is the center, there is no need to set the depth of search higher than 8 bit. For other initial points, many sweeps will be refused.

- The best initial point found is the center which mostly leads to the global maximum for all scenarios tested. However, if one wants to guarantee the global solution with any selected initial point as well as input scenarios, we refer to [49].
- Different values of the integer variables can lead to approximately the same objective function value when no curtailments are present and/or with slight differences in energy prices. For example, $t_1=7$, $t_2=14$, and $t_3=3$ (fixed method) and $t_1^{(*)}=8$, $t_2^{(*)}=16$, and $t_3^{(*)}=0$ (OTS method) can have a very slight difference in the objective function value, as shown in the case study.

Here, we summarize some advantages of our solution approach as follows:

1) The treatment of charging/discharging periods in the upper-stage makes the solution procedure highly effective.
2) Each iteration of the upper-stage is feasible, since all constraints are satisfied in the lower-stage.
3) Due to the limited search space of the upper-stage the number of iterations to reach the optimal solution will be low.

6.3 A Case study

The same medium-voltage network investigated in Section 5.2.3 is considered here as a case study. The on-, mid- and off-peak active energy prices are 117$/MWh, 100$/MWh and 65$/MWh from [124][42], respectively, as shown in Fig. 6.2. If the DSO takes hourly spot market prices instead of the fixed tariffs, hour-by-hour prices can be adopted and used to show how the effects if such prices are taken [59]. Generally, based on the TOU pricing, two to three price periods each day correspond with a good fit to an hour-by-hour prices as shown in [59] and [17]. This feature is relevant, since it is simpler for consumers to treat three prices than 24 different hourly prices each day. However, it can be desirable to update 24 different hourly prices tracking the expected spot prices [17] and gaining additional advantages. Therefore, a further hour-by-hour tariff is adapted around the fixed tariffs to show the impact of such prices on the performance of the system. Hourly prices assumed in winter and spring are given in Appendix C, Table C.5. The fixed reactive energy price is assumed to be 12\$/Mvarh [42]. The total wind power and demand scenarios in 4 different days considered for the optimization are shown in Fig. 6.6(a) to Fig. 6.11(a).

The optimization results listed in Table 6.1, Table 6.2, and Table 6.3 show the total revenue obtained by three methods for different wind/demand scenarios in four different days (see Fig. 6.6 to Fig. 6.11). The three methods include the modified OTS as well as enumeration to search for a flexible operation strategy for charge and discharge hours of the BSSs, and the fixed operation strategy for the BSSs. The results in Table 6.1, Table 6.2, and Table 6.3 are from the two-tariff, three-tariff, and 24-hour-tariff price model, respectively.

Using the modified OTS method we mostly obtained the same maximum as by enumeration, but it took much less computation time. However, a suitable initialization is required by the OTS to avoid local hills. It is clearly seen from Table 6.1, Table 6.2, and Table 6.3 that the total revenue based on the three-tariff and 24-hour-tariff price models are lower than that based on the two-tariff model. This is because the prices considered for the two-tariff price model are higher than the others, as shown in Fig. 6.2. It should be noted that the results here obtained are based on the condition of one cycle per day for the BSSs.

Table 6.1: Objective function value and computation time for different days in three methods using the two-tariff price model

	Modified OTS			Enumeration		Fixed A-R-OPF			Difference ($/day)
	Flexible $t_1^{(*)}$-$t_2^{(*)}$-$t_3^{(*)}$	CPU (sec.)	F ($/day)	CPU (sec.)	F ($/day)	Fixed t_1-t_2-t_3	CPU (sec.)	F ($/day)	
Day1	7-4-13	741	24587	7251	24587	7-14-3	26	24139	448(1.85%)
Day2	7-14-3	738	7620	7572	7620	7-14-3	28	7620	0(0%)
Day3	7-6-11	747	25815	7606	25815	7-14-3	33	25505	310(1.21%)
Day4	7-6-11	833	29593	7715	29593	7-14-3	30	29097	496(1.70%)

Table 6.2: Objective function value and computation time for different days in three methods using the three-tariff price model

	Modified OTS			Enumeration		Fixed A-R-OPF			Difference ($/day)
	Flexible $t_1^{(*)}$-$t_2^{(*)}$-$t_3^{(*)}$	CPU (sec.)	F ($/day)	CPU (sec.)	F ($/day)	Fixed t_1-t_2-t_3	CPU (sec.)	F ($/day)	
Day1	7-4-13	735	23088	6976	23088	7-14-3	25	22636	452(1.99%)
Day2	7-14-3	639	7481	6777	7481	7-14-3	23	7481	0(0%)
Day3	7-6-11	674	24332	7147	24332	7-14-3	33	24029	303(1.26%)
Day4	7-6-11	748	28014	7400	28014	7-14-3	33	27532	482(1.75%)

Table 6.3: Objective function value and computation time for different days in three methods using the 24-hour-tariff price model

	Modified OTS			Enumeration		Fixed A-R-OPF			Difference ($/day)
	Flexible $t_1^{(*)}$-$t_2^{(*)}$-$t_3^{(*)}$	CPU (sec.)	F ($/day)	CPU (sec.)	F ($/day)	Fixed t_1-t_2-t_3	CPU (sec.)	F ($/day)	
Day1	15-3-6	638	17701	6132	17701	7-14-3	19	17333	368(2.12%)
Day2	8-16-0	682	5647[1]	5894	5647[1]	7-14-3	23	5647[1]	0(0%)
Day3	14-5-5	720	18273	4623[2]	18273	7-14-3	22	18055	218(1.21%)
Day4	5-8-11	816	18386	6495	18386	7-14-3	18	18068	318(1.76%)

[1] There is a very slight difference, i.e., 5647.144244 ($/day) (fixed), 5647.145829 ($/day) (enumeration), and 5647.144248 ($/day) (OTS).
[2] The modified OTS finds a local solution, while the global maximum found by the enumeration method is 18285 $/day at $t_1^{(*)} = 7$, $t_2^{(*)} = 5$, $t_3^{(*)} = 12$.

By comparing the revenue differences between the flexible and fixed operation strategies, it can be seen that considerably more revenues can be gained in each day when high wind power generation occurs during midday hours. This takes place in the first, third, and fourth day, respectively.

In contrast, no difference can be seen in the second day, because the wind power generation occurs mostly during night hours. It is also noted that the integer variable t_1 does not change in all scenarios of the two-/three-tariff price model, because the energy prices are always low and constant during the first seven hours of the four days under consideration.

Note that the integer variable t_3 changes even with high energy prices before 21h due to high wind power generation and possible curtailments. Moreover, t_1 changes, as given in Table 6.3, when hourly prices are taken. This reflects the fact that the flexible strategy can achieve more revenue under different price models.

In addition, there is either no or with a very small difference (0.06%, see the third day in Table 6.3) between the enumeration method and the modified OTS. It was shown in Section 5.2 that a longer optimization horizon (with deterministic wind and demand power profiles) leads to more benefits not only economically but also in the sense of a smoothing operation. However, with a longer optimization period, uncertainty due to wind-power production and demand may increase.

Therefore, we solve the optimization problem for the four days with two different time horizons, namely 24h (procedure A) and 96h (procedure B), based on the three price models, respectively. The optimal trajectories are shown in Fig. 6.6 to Fig. 6.11, while the total revenue is given in Table 6.4. It is also seen that the revenue differences in procedure B are also considerable. In our study in Chapter 5, no reactive energy costs have been considered in the formulation of the A-R-OPF.

Now, to show the impact of the flexible operation strategy on the reactive energy import as well as cost from the connecting TN, we calculate the reactive energy cost by $\Sigma_h Q_{S1}(h) \times 12\$/$Mvarh. Here, $Q_{S1}(h)$ is the reactive power injected at slack bus in Mvar during hour h.

It is shown in Table 6.5 that the flexible operation strategy leads mostly to a higher cost compared with the fixed A-R-OPF. By comparing both differences, namely Diff.(P) and Diff.(Q), it can be clearly seen from Table 6.4 and Table 6.5 that the total gain from the flexible strategy dominates the total loss from the fixed strategy.

Table 6.4: Objective function value using the two-/three-/24-hour-tariff price model for procedure B in two methods

	Flexible A-R-OPF	Fixed A-R-OPF	Diff.(P) $/4-days
two-tariff $/4-days	87769	86471	+1298(+**1.50%**)
three-tariff $/4-days	83055	81778	+1277(+**1.56%**)
24-hour-tariff $/4-days	60165	59247	+918(+**1.55%**)

Table 6.5: Total cost of reactive energy import using the two-/three-/24-hour-tariff price model for procedure B in two methods

	Flexible A-R-OPF	Fixed A-R-OPF	Diff.(Q) $/4-days
two-tariff $/4-days	406	321	-85(-**26.48%**)
three-tariff $/4-days	427	419	-8(-**1.91%**)
24-hour-tariff $/4-days	-124	-114	+10(+**8.77%**)

Table 6.6: Total revenue from wind power and BSSs using the two-/three-/24-hour-tariff price model for procedure B in two methods

	Flexible A-R-OPF	Fixed A-R-OPF	Diff.(Revenue) $/4-days
two-tariff $/4-days	90147	88789	+1358(+**1.52%**)
three-tariff $/4-days	85281	83965	+1316(+**1.57%**)
24-hour-tariff $/4-days	61814	60851	+963(+**1.58%**)

Table 6.7: Total cost of energy losses using the two-/three-/24-hour-tariff price model for procedure B in two methods

	Flexible A-R-OPF	Fixed A-R-OPF	Diff.(Losses) $/4-days
two-tariff $/4-days	2378	2318	-60(-**2.59%**)
three-tariff $/4-days	2226	2187	-39(-**1.78%**)
24-hour-tariff $/4-days	1649	1604	-45(-**2.80%**)

Another point to note is that when using the three-/24-hour-tariff price model this impact will be either negligible (in the three-tariff) or even in the opposite direction (in the 24-hour-tariff). This is because the fixed strategy follows the energy prices during the charge/discharge process and has less attention on the reactive power. This can be clearly seen by comparing the differences during the third and fourth day in Fig. 6.6(b)-(d) to Fig. 6.11(b)-(d).

Since the objective function in this chapter has two main terms, it is useful to show their values separately. The first term is the total revenue from wind power and BSSs, as given in Table 6.6. It is clearly seen that this term is in accordance with the results in Table 6.4 for the two-/three-/24-hour-tariff price model. This is because the major saving comes from avoiding wind power curtailments through the flexible strategy. The second term is the total cost of energy losses, as given in Table 6.7. This term is in accordance with the results in Table 6.5 for the two-/three-/24-hour-tariff price model. This reflects the relationship between the power losses and reactive power flow in the network. We can conclude from these findings that more revenues can be obtained from the flexible A-R-OPF even with higher power losses.

Figs. 6.6(b)-(d) and 6.7(b)-(d) show the optimal profiles caused by the flexible (solid) and fixed A-R-OPF (dashed) obtained from procedure B using the two-tariff price model. It can be seen from Fig. 6.6(b)-(c) that the flexible strategy shifts the optimal control profiles of active power charge (positive part)/discharge (negative part) and reactive power dispatch. This leads to less wind power curtailments as seen in Fig. 6.6(d), which can be clearly seen in the first, third, and fourth day. Since there are three wind parks in the case study, the total curtailment factor axis has a range of (2.2–3), as seen e.g., in Fig. 6.6(d). In contrast, no curtailments are present in the second day due to low wind power generation. Fig. 6.7(b)-(c) shows the active and reactive power exchange at the slack bus. It can be seen that the flexible strategy leads to more active energy export or less energy import. However, the reactive energy import is seen in an opposite direction in comparison to the slack active power. This is because the active power charge/discharge of BSSs dominates the reactive power capability of the BSSs.

Fig. 6.7(d) shows large differences in the total energy level in the BSSs (i.e., the sum of the energy content of all 5 BSSs), especially in days when a high penetration of wind power generation is present. Similarly, Figs. 6.8(b)-(d) and 6.9(b)-(d) show the optimal trajectories of the flexible (solid) and fixed A-R-OPF (dashed) obtained from procedure B using the three-tariff price model. The same discussions presented above for the two-tariff price model are true for the three-tariff price model. However, in this case the flexible operation of BSSs avoids more wind power curtailments (see Fig. 6.8(d)), compared with that by the two-tariff price model. Figs. 6.10(b)-(d) and 6.11(b)-(d) show the impact if hourly prices are used instead of the fixed price models.

Since the difference between the highest and lowest active energy prices of $C_{pr.p}^{(C)}$, as shown in Fig. 6.2, is lower than those used in the two-/three-tariff price models, this leads to make more charge from wind power than from importing energy from the TN. This can be clearly seen in the first day where no charge occurs in the first 11 hours in comparison to the same period in Figs. 6.6(b) and 6.8(b). It means that the charge/discharge cycle is not always full-load, but it depends on input scenarios. However, as a rule for many practical applications, a cycle is considered to be full-load even if the storage system is not always used with its full capacity [114].

It is worth mentioning to note here that a price difference is required before arbitrage is performed for active power charge/discharge. For example, if there is no wind power generation in the considered four days, and the off-peak price is 65$/MWh in the two-tariff price model, the on-peak price should be at least 107$/MWh for the BSSs to begin to response for active power charge and discharge (see Fig. 6.2). Another important aspect is that the BSSs response always to reactive power dispatch of BSSs, even if the on-peak price is equal to off-peak price.

6.4 Conclusions

In this chapter, we have proposed a flexible battery management system (FBMS) for the operations of DNs with renewable penetration. In particular, the optimal lengths of charge/discharge cycle of BSSs for daily operations or even multiple days can lead to a considerably higher revenue in comparison to that from a fixed operation strategy. In addition, three different energy price models have been used and their impacts on the flexible operation compared. To solve the complex MINLP problem, we have proposed to separately treat the integer and continuous optimization variables, leading to a two-stage framework.

A real DN including dispersed wind parks, BSSs, and demands has been used as a case study. The effectiveness of the proposed FBMS is demonstrated through applying and testing different daily scenarios. It can be concluded that the proposed flexible and adaptive operation strategy will be promising for operating energy storage systems in the future energy networks.

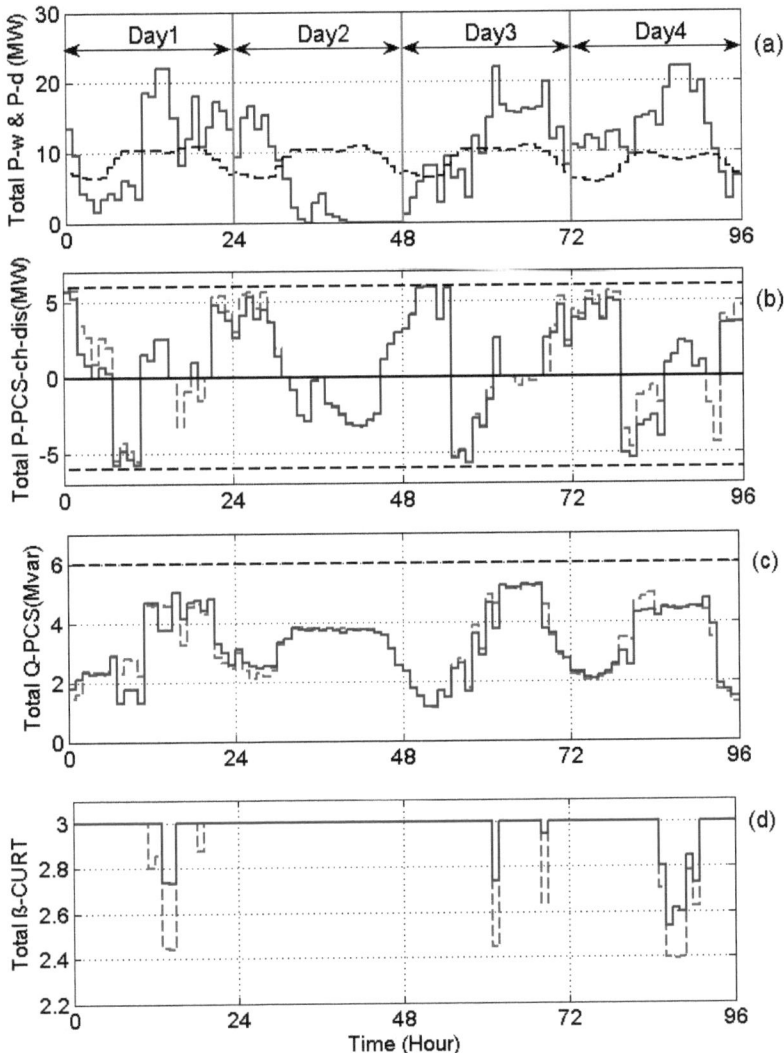

Figure 6.6: Trajectories by flexible and fixed A-R-OPF based on the two-tariff price model. (a) Total wind generation (solid-blue) and total demand power (dashed-black). (b) Total active power charge/discharge. (c) Total reactive power dispatch. (d) Total curtailment factor. Note: from (b) to (d) the lines (dashed-red) for fixed and (solid-blue) for flexible A-R-OPF.

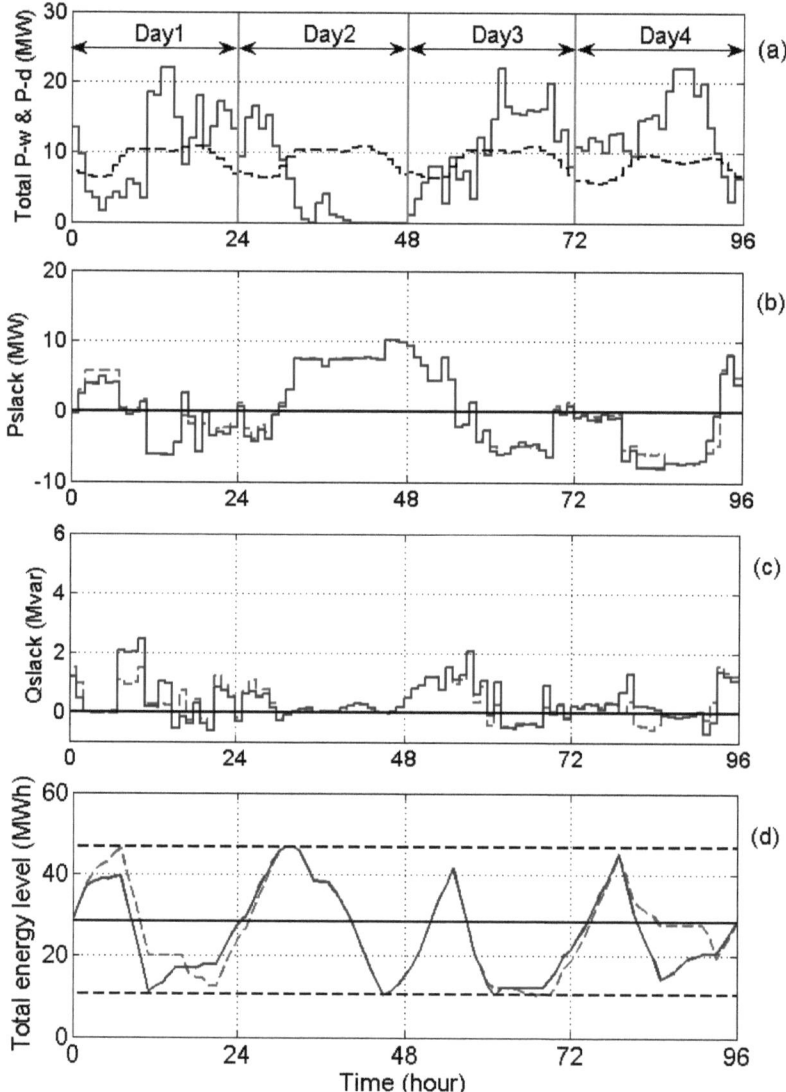

Figure 6.7: Trajectories by flexible and fixed A-R-OPF based on the two-tariff price model. (a) Total wind power generation (solid-blue) and total demand power (dashed-black). (b) Slack bus active power. (c) Slack bus reactive power. (d) Total energy level. Note: from (b) to (d) the lines (dashed-red) for fixed and (solid-blue) for flexible A-R-OPF.

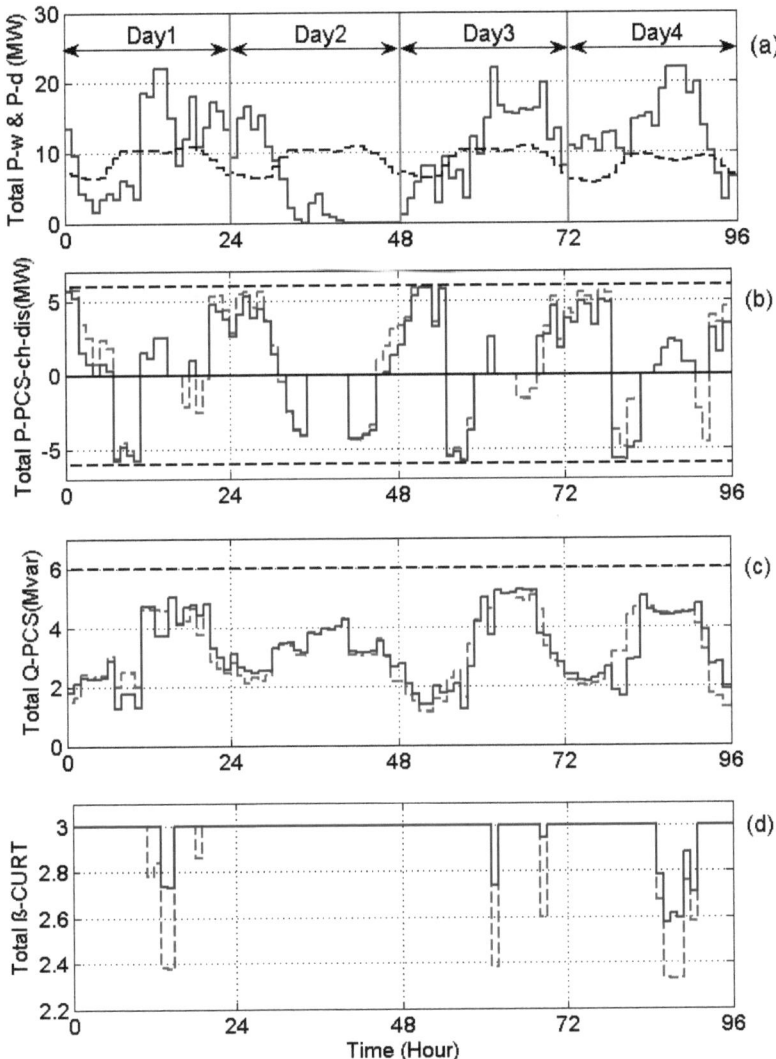

Figure 6.8: Trajectories by flexible and fixed A-R-OPF based on the three-tariff price model. (a) Total wind power generation (solid-blue) and total demand power (dashed-black). (b) Total active power charge/discharge. (c) Total reactive power dispatch. (d) Total curtailment factor. Note: from (b) to (d) the lines (dashed-red) for fixed and (solid-blue) for flexible A-R-OPF.

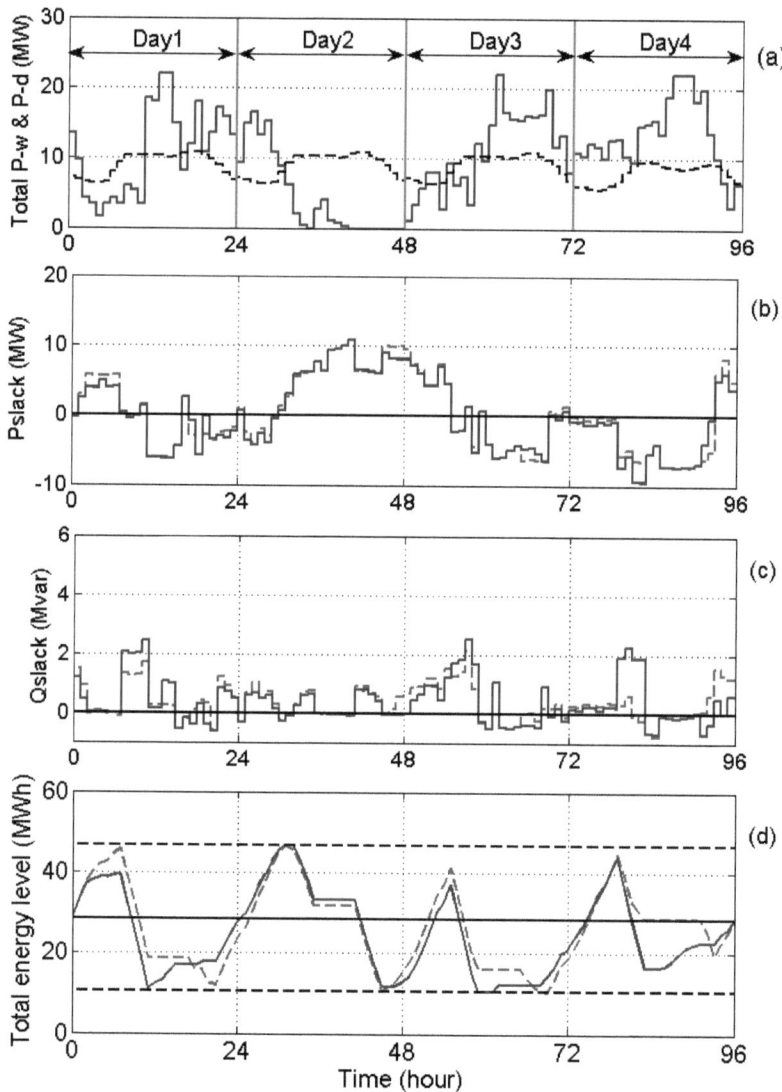

Figure 6.9: Trajectories by flexible and fixed A-R-OPF based on the three-tariff price model. (a) Total wind power generation (solid-blue) and total demand power (dashed-black). (b) Slack bus active power. (c) Slack bus reactive power. (d) Total energy level. Note: from (b) to (d) the lines (dashed-red) for fixed and (solid-blue) for flexible A-R-OPF.

Chap. 6: Flexible Optimal Operation of Battery Storage Systems for Energy Supply Networks

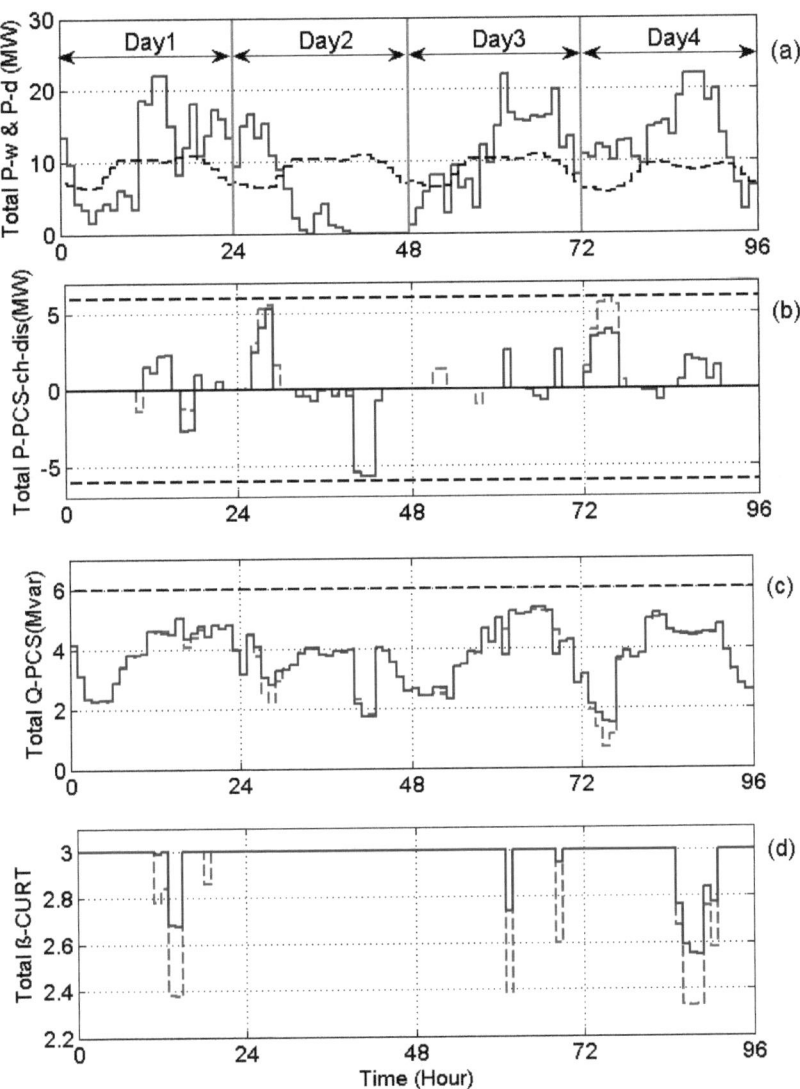

Figure 6.10: Trajectories by flexible and fixed A-R-OPF based on the 24-hour-tariff price model. (a) Total wind power generation (solid-blue) and total demand power (dashed-black). (b) Total active power charge/discharge. (c) Total reactive power dispatch. (d) Total curtailment factor. Note: from (b) to (d) the lines (dashed-red) for fixed and (solid-blue) for flexible A-R-OPF.

121

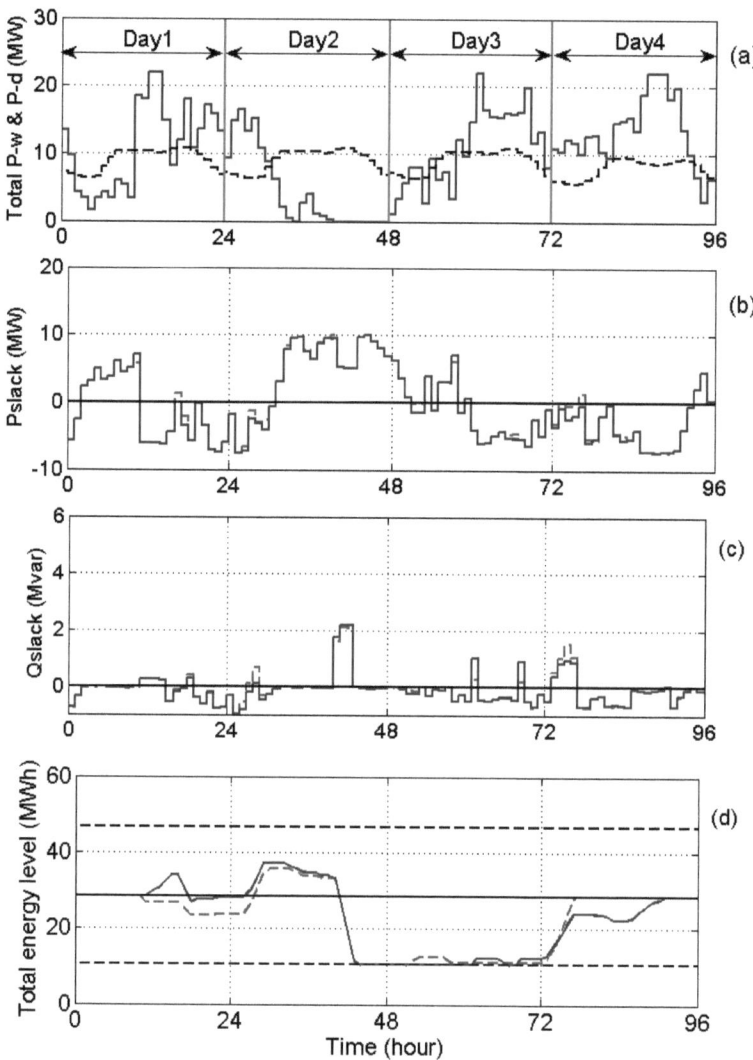

Figure 6.11: Trajectories by flexible and fixed A-R-OPF based on the 24-hour-tariff price model. (a) Total wind power generation (solid-blue) and total demand power (dashed-black). (b) Slack bus active power. (c) Slack bus reactive power. (d) Total energy level. Note: from (b) to (d) the lines (dashed-red) for fixed and (solid-blue) for flexible A-R-OPF.

7 Summary and Future Research Aspects

It is believed that renewable energies, e.g., wind and solar, will play an important role in the near future. This is because of its green nature in comparison to other conventional energy sources such as coal and oil. However, these renewable sources are highly fluctuating and not always available. For this reason, solutions have been recently proposed to reduce the effects of such variability and improve power system adequacy. A solution, which has been considered in the last decade, is the use of available energy storage systems, e.g., water pumped storage plants. Such plants are already built and being utilized in many countries. In addition, it may not be possible to construct new water storage plants because of less available water sources or geographical constraints. Therefore, other types of energy storage systems need to be further explored and exploited.

One of the promising solutions, considered for future power systems, is battery storage systems. Such systems can be integrated either in large scales on medium-voltage distribution networks or in small scales on low-voltage distribution networks. Currently, battery storage systems are relatively expensive, and therefore, many investigations have been carried out to analyze its costs and benefits. Despite its costs, it was shown that many advantages can be obtained from batteries if they are utilized.

From another perspective, already existing power networks can absorb a large amount of renewable energies. However, congestions or bottlenecks may occur in power systems because of line or feeder constraints. This is because both active and reactive powers are flowing simultaneously in power networks. Thus, if reactive power can be produced locally, i.e., near final consumers in distribution networks, then further space in power lines or feeders can be utilized to transport active power. Consequently, a large amount of renewable energies can be also accommodated by power networks without incurring additional costs, e.g., to upgrade network lines.

In this dissertation, we proposed an active-reactive optimal power flow method to explore the potential of power conditioning systems used typically to connect renewable energy generators and battery storage systems to distribution networks.

We analyzed two realistic case studies for two different voltage levels. For the low-voltage network, we consider the scenario where a high penetration of photovoltaic systems is implemented. An optimization problem is formulated with the aim to minimize the total cost of active energy losses and at the same time to minimize the total PV power curtailments. It was shown by utilizing the reactive power capability of photovoltaic system that there is no need to use batteries in low-voltage networks to save possible photovoltaic curtailments. This is because no further curtailments were observed. Moreover, unexpected high amounts of active energy losses arise in the grid because of a high difference in energy prices. Furthermore, a large amount of reactive energy is needed to be imported from an upstream connecting medium-voltage network.

In contrast, on the medium-voltage level, we considered a large amount of wind-based renewable energy generation to be accommodated with batteries. In this scenario, utilizing the reactive power capability of battery power conditioning systems has led to a significant reduction in energy losses and reactive energy needed to be imported from an upstream connecting transmission network. In contrast to the operation state in the studied low-voltage network, exports of reactive energy occurred in the medium-voltage network.

The strategy used for optimizing the operation of the medium-voltage network considered only one charge and discharge cycle every day. Since wind power, demand and energy prices vary from day to day, we proposed a flexible optimal operation strategy. In this strategy, higher revenues were obtained while maintaining the life of batteries. This can be made by considering one flexible charge/discharge cycle per day.

The operation of on-load-tap-changers of main transformers in distribution networks was also optimized to further reduce active energy losses in distribution networks without embedded generation units and batteries. The medium-voltage network is considered as a case study. As a result, a huge reduction of grid energy losses was obtained.

The above studies were made using software (simulators and optimizers) for power flow studies. This software has been developed as a part of the research work.

The future research aspects related to this work can be summarized as follows:

- The developed active and reactive power flow method in this work can be effectively used for planning distribution power systems considering both active and reactive power flow simultaneously.
- The impacts of bidirectional active and reactive power flow in connected power systems are important in terms of power system planning and operation. These impacts need to be further investigated.
- Economical aspects considering different energy prices and pricing mechanisms are concerns for future power grids.
- Uncertainties of renewable energies, demand as well as energy prices are additional challenges to be addressed in future works.

Appendix A: IEEE-RTS load data

Table A.1: Load data of the low- and medium-voltage DNs (Hourly demand as a percentage of the annual peak demand) [50][7]

Hour	Winter	Spring	Summer	Fall
1	0.4757	0.3969	0.64	0.3717
2	0.4473	0.3906	0.6	0.3658
3	0.426	0.378	0.58	0.354
4	0.4189	0.3654	0.56	0.3422
5	0.4189	0.3717	0.56	0.3481
6	0.426	0.4095	0.58	0.3835
7	0.5254	0.4536	0.64	0.4248
8	0.6106	0.5355	0.76	0.5015
9	0.6745	0.5985	0.87	0.5605
10	0.6816	0.6237	0.95	0.5841
11	0.6816	0.63	0.99	0.59
12	0.6745	0.6237	1	0.5841
13	0.6745	0.5859	0.99	0.5487
14	0.6745	0.5796	1	0.5428
15	0.6603	0.567	1	0.531
16	0.6674	0.5544	0.97	0.5192
17	0.7029	0.567	0.96	0.531
18	0.71	0.5796	0.96	0.5428
19	0.71	0.6048	0.93	0.5664
20	0.6816	0.6174	0.92	0.5782
21	0.6461	0.6048	0.92	0.5664
22	0.5893	0.567	0.93	0.531
23	0.5183	0.504	0.87	0.472
24	0.4473	0.441	0.72	0.413

Appendix B: Data for the low-voltage DN

Table B.1: Data of the low-voltage DN [4]

No. Line	From Bus	To Bus	Length (km)	R_l (ohm/km)	X_l (ohm/km)
1	1	2	0.100	0.195	0.070
2	2	3	0.137	1.900	0.100
3	3	4	0.168	1.900	0.100
4	4	5	0.010	1.900	0.100
5	2	6	0.107	1.900	0.100
6	6	7	0.102	1.900	0.100
7	2	8	0.162	0.868	0.078
8	8	9	0.081	0.383	0.101
9	9	10	0.070	0.868	0.078
10	10	11	0.093	0.868	0.078
11	11	12	0.174	1.117	0.410
12	11	13	0.066	0.868	0.078
13	13	14	0.086	0.868	0.078
14	14	15	0.173	0.868	0.078
15	14	16	0.104	0.195	0.070
16	10	17	0.073	1.117	0.410
17	17	18	0.119	0.519	0.350
18	18	19	0.145	1.117	0.410
19	19	20	0.041	1.900	0.100
20	18	21	0.067	0.519	0.350
21	21	22	0.121	0.519	0.350
22	22	23	0.119	0.519	0.350
23	23	24	0.036	0.868	0.078
24	23	25	0.100	0.868	0.078
25	22	26	0.149	1.117	0.410
26	18	27	0.049	0.519	0.350
27	9	28	0.035	1.900	0.100
28	8	29	0.084	1.900	0.100

Appendix B

Table B.2: Data of the low-voltage DN (Demand daily peak and PFs) [41]

Bus	$P_{peak}(i)$(kW)	PF	Bus	$P_{peak}(i)$(kW)	PF
3	1	0.9	17	5	0.9
4	3	0.9	20	3	0.9
5	16	0.9	25	3	0.9
6	3	0.9	26	3	0.9
7	1	0.9	27	3	0.9
13	3	0.9	29	3	0.9

Table B.3: Data of the low-voltage ADN (Upper bounds of active and reactive power in forward and reverse direction at slack bus) [41]

	$S_{S1.max} = 75$ kVA			
	$\alpha_{P1.fw}$	$\alpha_{Q1.fw}$	$\alpha_{P1.rev}$	$\alpha_{Q1.rev}$
Value	1	1	0.6	0.6

Table B.4: Data of the low-voltage ADN (PVSs) [41]

Bus	$S_{PCS.max.pv}(i)$ (kVA) = $P_{PV}(i)$ (kW)	Bus	$S_{PCS.max.pv}(i)$ (kVA) = $P_{PV}(i)$ (kW)
3	9	17	9
4	9	20	9
5	9	25	9
6	9	26	9
7	9	27	9
13	9	29	9

Appendix C: Data for the medium-voltage DN

Table C.1: Data of the medium-voltage DN [5]

No. Line	From Bus	To Bus	Length (km)	R_1 (ohm/km)	X_1 (ohm/km)	B_1 (µs/km)
1	1	2	5.7000	0.169111	0.418206	3.9540
2	2	3	1.0100	0.169111	0.418206	3.9540
3	2	4	0.4000	0.169111	0.418206	3.9540
4	4	5	0.3800	0.169111	0.418206	3.9540
5	5	6	0.1300	0.169111	0.418206	3.9540
6	5	7	0.1700	0.169111	0.418206	3.9540
7	7	9	0.2600	0.169111	0.418206	3.9540
8	9	10	0.1400	0.169111	0.418206	3.9540
9	9	11	0.3800	0.169111	0.418206	3.9540
10	11	12	0.5600	0.169111	0.418206	3.9540
11	12	13	0.3000	0.169111	0.418206	3.9540
12	12	14	3.3300	0.169111	0.418206	3.9540
13	14	15	1.0300	0.169111	0.418206	3.9540
14	16	17	1.0800	0.169111	0.418206	3.9540
15	17	18	1.6400	0.169111	0.418206	3.9540
16	18	19	0.4700	0.169111	0.418206	3.9540
17	19	20	0.4700	0.348124	0.468482	3.7571
18	21	22	0.9600	1.391924	0.478811	3.5971
19	19	23	0.1900	0.348124	0.468482	3.7571
20	23	24	1.9400	0.348124	0.468482	3.7571
21	24	25	2.4500	0.348124	0.468482	3.7571
22	24	26	1.6300	0.348124	0.468482	3.7571
23	26	27	1.2000	0.552276	0.485241	3.6035
24	26	28	2.1200	0.348124	0.468482	3.7571
25	28	29	0.7300	0.552276	0.485241	3.6035
26	29	30	0.7500	0.552276	0.485241	3.6035
27	28	31	2.5400	0.348124	0.468482	3.7571
28	23	32	0.3600	0.276519	0.458580	3.8280
29	32	33	0.2600	0.276519	0.458580	3.8280
30	33	34	3.5800	0.552276	0.485241	3.6035
31	33	35	0.7700	0.276519	0.458580	3.8280
32	35	36	2.0800	0.348124	0.468482	3.7571
33	35	37	4.5100	0.276519	0.458580	3.8280
34	37	38	3.2400	0.169111	0.418206	3.9540
35	38	39	0.3000	0.169111	0.418206	3.9540
36	39	40	0.5000	0.169111	0.418206	3.9540

Appendix C

Table C.2: Data of the medium-voltage DN (Demand daily peak and PFs) [40]

Bus	$P_{peak}(i)$	PF	Bus	$P_{peak}(i)$	PF
4	0.641346	0.95	25	0.028975	0.95
6	0.089706	0.87	27	0.015200	0.95
8	0.318725	0.95	30	0.019475	0.95
10	0.057600	0.75	31	0.051775	0.95
13	0.001900	1.00	34	0.020425	0.95
14	0.034675	0.95	36	0.008075	0.95
22	0.004750	0.95	37	0.010450	0.95
23	0.000950	0.95	41	0.216600	0.95

Table C.3: Data of the medium-voltage ADN (Upper bounds of active and reactive power in forward and reverse direction at slack bus) [40]

	$S_{S1.max}$ = 20 MVA			
	$\alpha_{P1.fw}$	$\alpha_{Q1.fw}$	$\alpha_{P1.rev}$	$\alpha_{Q1.rev}$
Value	1	1	0.6	0.6

Table C.4: Data of the medium-voltage ADN (wind turbines, PCSs capabilities and BSSs capacities) [40]

	BSSs stations			Wind-BSSs stations		
Bus	4	9	39	19	28	40
P_W	-	-	-	0.8	0.4	1
$S_{PCS.max.b}$	0.2	0.15	0.1	-	0.05	0.1
E_{BSS}	1.948	1.299	0.455	-	0.844	0.649

Table C.5: Data of the medium-voltage DN (active energy prices for 24-hour-tariff price model in winter and spring) [42]

h	Price ($/MWh)		h	Price ($/MWh)		h	Price ($/MWh)	
	winter	spring		winter	spring		winter	spring
1	55.65	46.43	9	78.91	70.02	17	82.23	66.33
2	52.33	45.70	10	79.47	72.97	18	83.07	67.81
3	49.84	44.22	11	79.47	73.71	19	83.07	70.76
4	49.01	42.75	12	78.91	72.97	20	79.74	72.23
5	49.01	43.48	13	78.91	68.55	21	75.59	70.76
6	49.84	47.91	14	78.91	67.81	22	68.94	66.33
7	61.47	53.07	15	77.25	66.33	23	60.64	58.96
8	71.44	62.65	16	78.08	64.86	24	52.33	51.59

Appendix D

Appendix D: Software implementation of DSI-1

Here is the implementation of the DSI-1 used for carrying out dynamic power flow studies in DNs. The DSI-1 is implemented in the MATLAB-Simulink environment with user-interfaces, as shown in Figs. D.1 and D.2. It is basically a hierarchical model comprising many layers and subsystems.

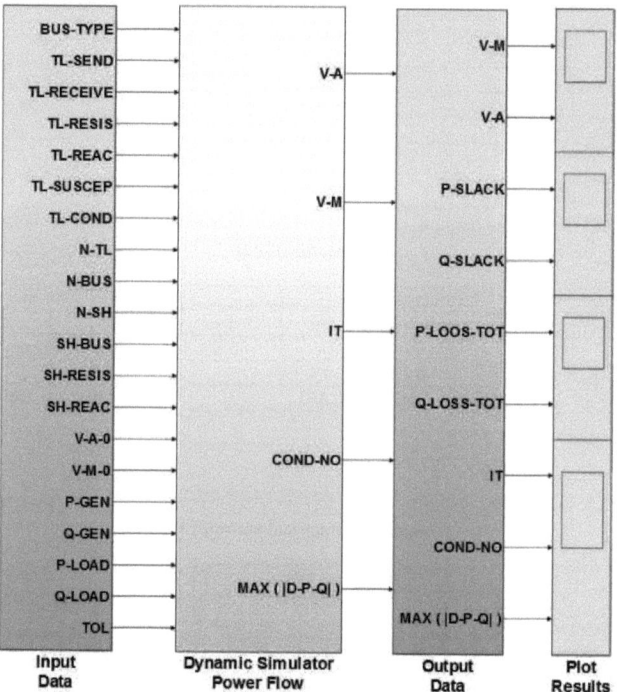

Figure D.1: Main user-interface of the DSI-1 in the MATLAB-Simulink environment.

Appendix D

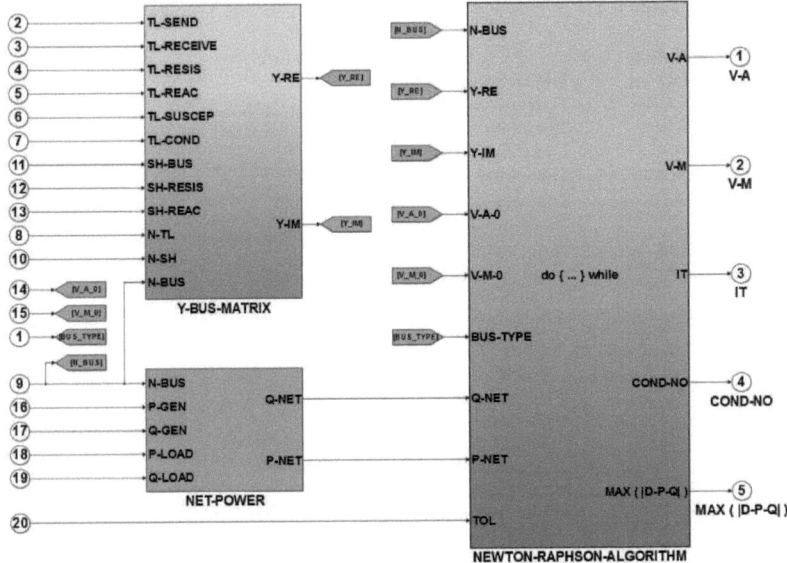

Figure D.2: Schema of the Dynamic Simulator block in Fig. D.1.

Appendix D

Table D.1: Symbols and descriptions of DSI-1 in Figs. D.1 and D.2

Symbol	Description
TL-SEND	Transmission line (send bus)
TL-RECEIVE	Transmission line (receive bus)
TL-RESIS	Transmission line (resistance)
TL-REAC	Transmission line (reactance)
TL-SUSCEP	Transmission line (susceptance)
TL-COND	Transmission line (conductance)
SH-BUS	Shunt elements (number of buses)
SH-RESIS	Shunt elements (resistance)
SH-REAC	Shunt elements (reactance)
N-TL	Number of transmission lines
N-BUS	Number of buses
N-SH	Number of shunt elements
TOL	Tolerance of the calculation
BUS-TYPE	Bus type
IT	Number of iterations
COND-NO	Condition number of the Jacobian-Matrix
D-P-Q	Active-reactive power mismatch
V-M-0	Initial voltage amplitude
V-A-0	Initial voltage angle
V-M	Voltage amplitude
V-A	Voltage angle
P-LOAD	Load active power
Q-LOAD	Load reactive power
P-GEN	Generation active power
Q-GEN	Generation reactive power
Y-RE	Real part of the admittance matrix
Y-IM	Imaginary part of the admittance matrix
P-NET	Scheduled active power
Q-NET	Scheduled reactive power
P-LOOS-TOT	Total active power losses
Q-LOSS-TOT	Total reactive power losses
P-SLACK	Slack active power
Q-SLACK	Slack reactive power

Appendix E: Software implementation of DSI-2

Here is the implementation of the DSI-2 which is a main TR control system. This simulator is implemented in the MATLAB-Simulink environment with a user-interface, as shown in Fig. E.1.

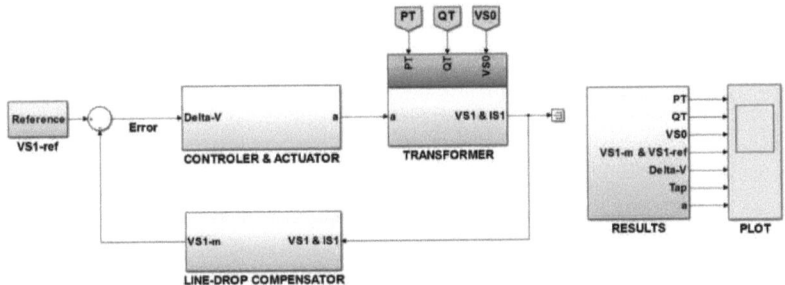

Figure E.1: Main user-interface of the DSI-2 in the MATLAB-Simulink environment.

Table E.1: Symbols and descriptions of DSI-2 in Fig. E.1

Symbol	Description
PT	Transformer active power load
QT	Transformer reactive power load
VS0	Primary transformer voltage
VS1-ref	Reference voltage
VS1	Secondary transformer voltage
IS1	Secondary transformer current
VS1-m	Measuring voltage after line-drop compensator voltage
Delta-V	Voltage error
Tap	Tap position
a	Transformer tap-ratio

Appendix D

Bibliography

[1] E. Acha, C. R. Fuerte-Esquivel, H. Ambriz-Perez, and C. Angeles-Camacho, *Facts-Modelling and Simulation in Power Networks*. NY: Wiley, 2004, pp. 98-111.

[2] P. R. Adby and M. A. Dempster, *Introduction to Optimization Methods*. London, U.K.: Chapman & Hall, 1974.

[3] K. -H. Ahlert, "Economics of distributed storage systems–an economic analysis of arbitrage-maximizing storage systems at the end consumer," Ph.D. dissertation, Dept. Economics and Business Eng., Karlsruhe Institute of Technology (KIT), 2010.

[4] R. Angelino, G. Carpinelli, D. Proto, and A. Bracale, "Dispersed generation and storage systems for providing ancillary services in distribution systems," *International Symposium on Power Electronics Electrical Drives Automation and Motion (SPEEDAM)*, pp. 343-351, 14-16 Jun. 2010.

[5] Y. M. Atwa and E. F. El-Saadany, "Probabilistic approach for optimal allocation of wind-based distributed generation in distribution systems," *IET Renewable Power Generation*, vol. 5, no. 1, pp. 79-88, Jan. 2011.

[6] Y. M. Atwa, E. F. El-Saadany, M. M. A. Salama, and R. Seethapathy, "Optimal renewable resources mix for distribution system energy loss minimization," *IEEE Transactions on Power Systems*, vol. 25, no. 1, pp. 360-370, Feb. 2010.

[7] Y. M. Atwa and E. F. El-Saadany, "Optimal allocation of ESS in distribution systems with a high penetration of wind energy," *IEEE Transactions on Power Systems*, vol. 25, no. 4, pp. 1815-1822, Nov. 2010.

[8] Y. M. Atwa, "Distribution system planning and reliability assessment under high DG penetration," Ph.D. dissertation, Dept. Electrical and Computer Eng., Waterloo, Ontario, Canada, 2010.

[9] A. G. Bakirtzis, P. N. Biskas, C. E. Zoumas, and V. Petridis, "Optimal power flow by enhanced genetic algorithm," *IEEE Trans. Power Syst.*, vol. 17, no. 2, pp. 229-236, May 2002.

[10] B. Borkowska, "Probabilistic load flow," *IEEE Transactions on Power Apparatus and Systems*, vol. PAS-93, no. 3, pp. 752-759, May 1974.

Bibliography

[11] L. J. Borle, M.S. Dymond, and C.V. Nayar, "Development and testing of a 20–kW grid interactive photovoltaic power conditioning system in Western Australia," *IEEE Transactions on Industry Applications*, vol. 23, no. 2, pp. 502-508, Mar./Apr. 1997.

[12] A. Bracale, R. Angelino, G. Carpinelli, M. Mangoni, and D. Proto, "Dispersed generation units providing system ancillary services in distribution networks by a centralised control," *IET Renewable Power Generation*, vol. 5, no. 4, pp. 311-321, Jul. 2011.

[13] W. H. Browne, "Power factor indicators," *Transactions of the American Institute of Electrical Engineers*, vol. XVIII, pp.287-312, Jan. 1901.

[14] A. Cagnano, E. De Tuglie, M. Liserre, and R.A. Mastromauro, "Online optimal reactive power control strategy of PV inverters," *IEEE Transactions on Industrial Electronics*, vol. 58, no. 10, pp. 4549-4558, Oct. 2011.

[15] M. S. Calovic, "Modeling and analysis of under-load tap-changing transformer control systems," *IEEE Transactions on Power Apparatus and Systems*, vol. PAS-103, no. 7, pp.1909-1915, Jul. 1984.

[16] C. A. Canizares, K. Bhattacharya, I. El-Samahy, H. Haghighat, J. Pan, and C. Tang, "Re-defining the reactive power dispatch problem in the context of competitive electricity markets," *IET Gen., Transm. Distrib.*, vol. 4, iss. 2, pp. 162-177, Feb. 2010.

[17] M. C. Caramanis, R.E. Bohn, and F.C. Schweppe, "Optimal spot pricing: practice and theory," *IEEE Transactions on Power Apparatus and Systems*, vol. PAS-101, no. 9, pp. 3234-3245, Sep. 1982.

[18] J. Carpentier, "Contribution to the economic dispatch problem," *Bull. Soc. Franc. Elect.*, vol. 8, pp. 431-447, 1962.

[19] R. Cecchini and G. Pelosi, "Alessandro Volta and his battery," *IEEE Antennas and Propagation Magazine*, vol. 34, no. 2, pp. 30-37, Apr. 1992.

[20] F. A. Chacra, P. Bastard, and G. Fleury, "Energy storage associated to wind farms in a two-tariff structure bidding tender," *IEEE Russia Power Tech*, pp. 1-5, 27-30 June 2005.

[21] F. A. Chacra, P. Bastard, G. Fleury, and R. Clavreul, "Impact of energy storage costs on economical performance in a distribution substation," *IEEE Transactions on Power Systems*, vol. 20, no. 2, pp. 684-691, May 2005.

[22] D. Chattopadhyay, "Application of general algebraic modeling system to power system optimization," *IEEE Transactions on Power Systems*, vol. 14, no. 1, pp.15-22, Feb. 1999.

[23] C. Chen, S. Duan, T. Cai, B. Liu, and G. Hu, "Optimal allocation and economic analysis of energy storage system in microgrids," *IEEE Transactions on Power Electronics*, vol. 26, no. 10, pp. 2762-2773, Oct. 2011.

[24] S. P. Chowdhury, P. Crossley, and S. Chowdhury, "Microgrids and active distribution networks," *IET Renewable Energy Series 6, The institution of engineering and technology*, pp. 1-13, 2009.

[25] E. V. Clark, "Power factor and tariff," *Journal of the Institution of Electrical Engineers*, vol. 64, no. 354, pp. 625-632, Jun. 1926.

[26] B. J. Davidson, I. Glendenning, R. D. Harman, A. B. Hart, B. J. Maddock, R. D. Moffitt, V. G. Newman, T. F. Smith, P. J. Worthington, and J. K. Wright, "Large-scale electrical energy storage," *Physical Science, Measurement and Instrumentation, Management and Education-Reviews, IEE Proceedings A*, vol. 127, no. 6, pp. 345-385, Jul. 1980.

[27] M. V. D.-Dobrowolsky, "Apparatus for indicating difference of phase." U.S. Patent 549,449, issued Nov. 5, 1895.

[28] H. W. Dommel and W.F. Tinney, "Optimal power flow solutions," *IEEE Transactions on Power Apparatus and Systems*, vol. PAS-87, no. 10, pp. 1866-1876, Oct. 1968.

[29] E. W. Dorey, "The improvement of power factor," *Journal of the Institution of Electrical Engineers*, vol. 64, no. 354, pp. 633-654, Jun. 1926.

[30] L. A. Dunstan, "Machine computation of power network performance," *Transactions of the American Institute of Electrical Engineers*, vol. 66, no. 1, pp. 610-624, Jan. 1947.

[31] H. Dura, "Optimum number, location, and size of shunt capacitors in radial distribution feeders a dynamic programming approach," *IEEE Transactions on Power Apparatus and Systems*, vol. PAS-87, no. 9, pp. 1769-1774, Sep. 1968.

[32] T. A. Edison, "Electric-Lamp." U.S. Patent 223,898, issued Jan. 27, 1880.

[33] T. A. Edison, "Electric meter." U.S. Patent 251,545, issued Dec. 27, 1881.

[34] T. A. Edison, "System of underground conductors for electrical distribution." U.S. Patent 273,828, issued Mar. 13, 1883.

[35] T. A. Edison, "System of electrical distribution." U.S. Patent 385,173, issued Jun. 26, 1888.

[36] M. E. El-Hawary, *Electrical Power systems: Design and Analysis*, Rev. ed. NY: IEEE PRESS, 1995.

Bibliography

[37] A. Faruqui, "Pricing programs: time-of-use and real time," *Encycl. Energy Eng. Technol.*, pp. 1175-1183, Feb. 2008.

[38] F. W. Felix, G. Gross, J. F. Luini, and P. M. Look, "A Two-stage approach to solving large-scale optimal power flows," *Power Industry Computer Applications Conference*, PICA-79. *IEEE Conference Proceedings*, pp. 126-136, 15-18 May 1979.

[39] A. Gabash and P. Li, "Evaluation of reactive power capability by optimal control of wind-vanadium redox battery stations in electricity market," *Renewable Energy & Power Quality Journal*, no. 9, pp. 1-6, May 2011.

[40] A. Gabash and P. Li, "Active-Reactive optimal power flow in distribution networks with embedded generation and battery storage," *IEEE Transaction on Power Systems*, vol. 27, no. 4, pp. 2026-2035, Nov. 2012.

[41] A. Gabash and P. Li, "Active-Reactive optimal power flow for low-voltage networks with photovoltaic distributed generation," *2nd IEEE International Energy Conference and Exhibition (EnergyCon2012)/ Future Energy Grids and Systems (FEGS)*, Florence, Italy, Sep. 2012, pp. 381-386.

[42] A. Gabash and P. Li, "Flexible optimal operation of battery storage systems for energy supply networks," *IEEE Transaction on Power Systems*, vol. 28, no. 3, pp. 2788-2797, Aug. 2013.

[43] A. Gabash, D. Xie, and P. Li, "Analysis of influence factors on rejected active power from active distribution networks," *Power & Energy Student Summit (PESS) 2012, IEEE Student Branch TU-Ilmenau*, Ilmenau, Germany, Jan. 2012, pp. 25-29.

[44] A. Gabash, M. E. Alkal, and P. Li, "Impact of allowed reverse active power flow on planning PVs and BSSs in distribution networks considering demand and EVs growth," *Power & Energy Student Summit (PESS) 2013, IEEE Student Branch Bielefeld*, Bielefeld, Germany, Jan. 2013, pp. 11-16.

[45] E. E. George, "Intrasystem transmission losses," *Transactions of the American Institute of Electrical Engineers*, vol. 62, no. 3, pp.153-158, Mar. 1943.

[46] F. Geth, J. Tant, E. Haesen, J. Driesen, and R. Belmans, "Integration of energy storage in distribution grids," *Power and Energy Society General Meeting, 2010 IEEE*, pp.1-6, Jul. 2010.

[47] F. Geth, J. Tant, T. De Rybel, P. Tant, D. Six, and J. Driesen, "Techno-economical and life expectancy modeling of battery energy storage systems," *21st International Conference and Exhibition on Electricity Distribution*, no. 1106, pp.1-4, Jun. 2011.

[48] A. F. Glimn and G. W. Stagg, "Automatic calculation of load flows," *Power Apparatus and Systems, Part III. Transactions of the American Institute of Electrical Engineers*, vol. 76, no. 3, pp. 817-825, Apr. 1957.

[49] D. E. Goldberg, *Genetic Algorithms in Search, Optimization and Machine Learning*. Reading, MI: Addison-Wesley, 1989.

[50] C. Grigg, P. Wong, P. Albrecht, R. Allan, M. Bhavaraju, R. Billinton, Q. Chen, C. Fong, S. Haddad, S. Kuruganty, W. Li, R. Mukerji, D. Patton, N. Rau, D. Reppen, A. Schneider, M. Shahidehpour, and C. Singh, "The IEEE reliability test system-1996. A report prepared by the reliability test system task force of the application of probability methods subcommittee," *IEEE Transactions on Power Systems*, vol. 14, no. 3, pp. 1010-1020, Aug. 1999.

[51] G. P. Harrison and A. R. Wallace, "Optimal power flow evaluation of distribution network capacity for the connection of distributed generation," *IEE Proceedings Generation, Transmission and Distribution*, vol. 152, no. 1, pp. 115-122, 10 Jan. 2005.

[52] H. L. Hazen, O. R. Schurig, and M. F. Gardner, "The M. I. T. network analyzer design and application to power system problems," *Transactions of the American Institute of Electrical Engineers*, vol. 49, no. 3, pp. 1102-1113, Jul. 1930.

[53] J. Hetzer, D.C. Yu, and K. Bhattarai, "An economic dispatch model incorporating wind power," *IEEE Transactions on Energy Conversion*, vol. 23, no. 2, pp. 603-611, Jun. 2008.

[54] K. Heuck, K.-D. Dettmann, and D. Schulz, *Elektrische Energieversorgung*, ed. 8, Springer: 2010.

[55] C. Higgins and E. W. W. Edwards, *Electric Lighting Act*. William Clowes & Sons. 1883, pp. 71-76.

[56] J. H. Holland, *Adaptive in Natural and Artificial Systems*. Ann Arbor, MI: Univ. Michigan Press, 1975.

[57] M. Huneault and F. D. Galiana, "A survey of the optimal power flow literature," *IEEE Transactions on Power systems*, vol. 6, no. 2, pp. 762-770, May 1991.

[58] B. Jansen, "Einrichtung zum Umschaltung zweier Anzapfungen eines Stufentransformers während des Betriebes durch zwei gegenläufig bewegte Leistungsschalter mit Vorkontakten," Deutsches Patent 474613, issued Jul. 13, 1926.

[59] P. M. D. O. -D. Jesus, M.T.P. de Leao, J.M. Yusta, H.M. Khodr, and A.J. Urdaneta, "Uniform marginal pricing for the remuneration of distribution networks," *IEEE Transactions on Power Systems*, vol. 20, no. 3, pp. 1302-1310, Aug. 2005.

[60] J. A. Johnson, "Operating aspects of reactive power," *Transactions of the American Institute of Electrical Engineers*, vol. 52, no. 3, pp. 752-757, Sep. 1933.

[61] C. H. Jolissaint, N.V. Arvanitidis, and D.G. Luenberger, "Decomposition of real and reactive power flows: A method suited for on-line applications," *IEEE Transactions on Power Apparatus and Systems*, vol. PAS-91, no. 2, pp. 661-670, March 1972.

[62] P. Jong-Young, S. Jin-Man, and P. Jong-Keun, "Optimal capacitor allocation in a distribution system considering operation costs," *IEEE Transactions on Power Systems*, vol. 24, no. 1, pp. 462-468, Feb. 2009.

[63] G. Kapp, "The improvement of power factor," *Journal of the Institution of Electrical Engineers*, vol. 61, no. 314, pp. 89-108, Jan. 1923.

[64] S. A. Kazarlis, A. G. Bakirtzis, and V. Petridis, "A genetic algorithm solution to the unit commitment problem," *IEEE Trans. Power Syst.*, vol. 11, no. 1, pp. 83-92, Feb. 1996.

[65] A. Keane, L. F. Ochoa, C. L. T. Borges, G. W. Ault, A. D. Alarcon-Rodriguez, R. Currie, F. Pilo, C. Dent, and G. P. Harrison, "State-of-the-art techniques and challenges ahead for distributed generation planning and optimization," *IEEE Transactions on Power Systems*, vol. 28, no. 2, pp.1493-1502, May 2013.

[66] J. Kennedy and R. Eberhart, "Particle swarm optimization," in *Proc. IEEE Int. Conf. Neural Networks*, 1995, vol. 4, pp. 1942-1948.

[67] J. Kennedy and R. C. Eberhart, "A discrete binary version of the particle swarm algorithm," in *Proc. Conf. Systems, Man, Cybern.*, Oct. 1997, pp. 4104-4109.

[68] W. H. Kersting, *Distribution System Modeling and Analysis*, 1st ed. Boca Raton, FL: CRC Press, 2007.

[69] R. J. Konopinski, P. Vijayan, and V. Ajjarapu, "Extended reactive capability of DFIG wind parks for enhanced system performance," *IEEE Transactions on Power Electronics*, vol. 24, no. 3, pp. 1346-1355, Aug. 2009.

[70] S. N. Liew and G. Strbac, "Maximising penetration of wind generation in existing distribution networks," *IEE Proceedings Generation, Transmission and Distribution*, vol. 149, no. 3, pp. 256-262, May 2002.

[71] W. E. Mabeea, J. Mannionb, and T. Carpenterc, "Comparing the feed-in tariff incentives for renewable electricity in Ontario and Germany," *Energy Policy*, 40(2012), pp. 480-489.

[72] C. MacKechnie Jarvis, "Nikola Tesla and the induction motor," *Electronics and Power*, vol. 15, no.12, pp. 436-440, Dec. 1969.

[73] A. S. Marincic, "Nikola Tesla and the wireless transmission of energy," *IEEE Transactions on Power Apparatus and Systems*, vol. PAS-101, no.10, pp. 4064-4068, Oct. 1982.

[74] N. W. Miller, R. S. Zrebiec, G. Hunt, and R. W. Deimerico, "Design and commissioning of a 5 MVA, 2.5 MWh battery energy storage system," *IEEE Proceedings. Transmission and Distribution Conference*, pp. 339-345, 15-20 Sep. 1996.

[75] J. A. Momoh, R. Adapa, and M. E. El-Hawary, "A review of selected optimal power flow literature to 1993. I. Nonlinear and quadratic programming approaches," *IEEE Transactions on Power Systems*, vol. 14, no. 1, pp. 96-104, Feb. 1999.

[76] J. A. Momoh, M. E. El-Hawary, and R. Adapa, "A review of selected optimal power flow literature to 1993. II. Newton, linear programming and interior point methods," *IEEE Transactions on Power Systems*, vol. 14, no. 1, pp. 105-111, Feb. 1999.

[77] Nikola Tesla 1857-1943, *Proceedings of the I.R.E.*, pp. 194, May 1943.

[78] L. F. Ochoa, C.J. Dent, and G.P. Harrison, "Distribution network capacity assessment: Variable DG and active networks," *IEEE Transactions on Power Systems*, vol. 25, no. 1, pp. 87-95, Feb. 2010.

[79] L. F. Ochoa and G.P. Harrison, "Minimizing energy losses: Optimal accommodation and smart operation of renewable distributed generation," *IEEE Transactions on Power Systems*, vol. 26, no. 1, pp. 198-205, Feb. 2011.

[80] L. F. Ochoa, A. Keane, and G.P. Harrison, "Minimizing the reactive support for distributed generation: Enhanced passive operation and smart distribution networks," *IEEE Transactions on Power Systems*, vol. 26, no. 4, pp. 2134-2142, Nov. 2011.

[81] H. Oh, "Optimal planning to include storage devices in power systems," *IEEE Transactions on Power Systems*, vol. 26, no. 3, pp. 1118-1128, Aug. 2011.

Bibliography

[82] P. M. D. Oliveira, P.M. Jesus, E.D. Castronuovo, and M.T. Leao, "Reactive power response of wind generators under an incremental network-loss allocation approach," *IEEE Transactions on Energy Conversion*, vol. 23, no. 2, pp. 612-621, Jun. 2008.

[83] J. Peschon, D. S. Piercy, W. F. Tinney, O. J. Tveit, and M. Cuenod, "Optimum control of reactive power flow," *IEEE Transactions on Power Apparatus and Systems*, vol. PAS-87, no. 1, pp. 40-48, Jan. 1968.

[84] J. Peschon, D.S. Piercy, W. F. Tinney, and O. J. Tveit, "Sensitivity in power systems," *IEEE Transactions on Power Apparatus and Systems*, vol. PAS-87, no. 8, pp.1687-1696, Aug. 1968.

[85] V. Petridis, S. Kazarlis, and A. Bakirtzis, "Varying fitness functions in genetic algorithm constrained optimization: The cutting stock and unit commitment problems," *IEEE Trans. Syst., Man, Cybern. B*, vol. 28, pp. 629-640, Oct. 1998.

[86] R. A. Philip, "The flow of energy," *Transactions of the American Institute of Electrical Engineers*, vol. XXXIV, no. 1, pp.779-808, Jan. 1915.

[87] P. Poonpun and W. T. Jewell, "Analysis of the cost per kilowatt hour to store electricity," *IEEE Transactions on Energy Conversion*, vol. 23, no. 2, pp. 529-534, Jun. 2008.

[88] N. S. Rau and Y. -H. Wan, "Optimum location of resources in distributed planning," *IEEE Transactions on Power Systems*, vol., no. 4, pp. 2014-2020, Nov. 1994.

[89] T. B. Reddy, *Linden's Handbook of Batteries*. 4th ed. NY: McGraw-Hill, 2010.

[90] P. F. Ribeiro, B. K. Johnson, M. L. Crow, A. Arsoy, and Y. Liu, "Energy storage systems for advanced power applications," *Proceedings of the IEEE*, vol. 89, no. 12, pp.1744-1756, Dec. 2001.

[91] A. L. Samuel, "Some studies in machine learning using the game of checkers," *IBM J. Res. Develop.*, vol. 3, no. 3, pp. 210-219, 1959.

[92] A. M. Sasson, "Nonlinear programming solutions for load-flow, minimum-loss, and economic dispatching problems," *IEEE Transactions on Power Apparatus and Systems*, vol. PAS-88, no. 4, pp. 399-409, Apr. 1969.

[93] S. M. Schoenung and W. V. Hassenzahl, "Long- vs. short-term energy storage technologies analysis: A life-cycle cost study," Sandia Natl. Lab., Albuquerque, NM, Sandia Rep. SAND2003-2783, 2003.

[94] F. C. Schweppe, M. C. Caramanis, R. D. Tabors, and R. E. Bohn, *Spot Pricing of Electricity*, Boston, MA: Kluwer, 1988.

[95] M. Shahidehpour, H. Yamin, and Z. Li, *Market Operation in Electric Power Systems. Forecasting, Scheduling, and Risk Management*. NY: Wiley-IEEE Press, 2002.

[96] R. A. Shayani and M.A.G. de Oliveira, "Photovoltaic generation penetration limits in radial distribution systems," *IEEE Transactions on Power Systems*, vol. 26, no. 3, pp. 1625-1631, Aug. 2011.

[97] R. R. Shoults and D. T. Sun, "Optimal power flow based upon P-Q decomposition," *IEEE Transactions on Power Apparatus and Systems*, vol. PAS-101, no. 2, pp. 397-405, Feb. 1982.

[98] H. M. Smith and S.-Y. Tong, "Minimizing power transmission losses by reactive-volt-ampere control," *IEEE Transactions on Power Apparatus and Systems*, vol. 82, no. 67, pp. 542-544, Aug. 1963.

[99] R. Srinivasan and K. R. Rao, "Predictive coding based on efficient motion estimation," *IEEE Trans. Commun.*, vol. COM-33, no. 8, pp. 888-896, Aug. 1985.

[100] W. Stanley, "Induction coil." U.S. Patent 349,611, issued Sep. 21, 1886.

[101] R. H. Stevens, "Power flow direction definitions for metering of bidirectional power," *IEEE Trans. Power App. Syst.*, vol. PAS-102, no. 9, pp. 3018-3022, Sep. 1983.

[102] J. Tant, F. Geth, D. Six, P. Tant, and J. Driesen, "Multiobjective battery storage to improve PV integration in residential distribution grids," *IEEE Transactions on Sustainable Energy*, vol. 4, no. 1, pp. 182-191, Jan. 2013.

[103] S. Teleke, M. E. Baran, A. Q. Huang, S. Bhattacharya, and L. Anderson, "Control strategies for battery energy storage for wind farm dispatching," *IEEE Transactions on Energy Conversion*, vol. 24, no. 3, pp. 725-732, Sep., 2009.

[104] S. Teleke, M. E. Baran, S. Bhattacharya, and A. Q. Huang, "Optimal control of battery energy storage for wind farm dispatching," *IEEE Transactions on Energy Conversion*, vol. 25, no. 3, pp. 787-794, Sep., 2010.

[105] A. Ter-Gazarian, *Energy Storage for Power Systems*, Peter Peregrinus Ltd., U.K.: London, 1994.

[106] N. Tesla, "Electro magnetic motor." U.S. Patent 381,968, issued May 1, 1888.

[107] N. Tesla, "Electrical transmission of power." U.S. Patent 382,280, issued May 1, 1888.

[108] The sign of reactive power, *Electrical Engineering*, vol. 65, no. 11, pp. 512-516, Nov. 1946.

Bibliography

[109] The sign of reactive power —II, *Electrical Engineering*, vol. 67, no. 1, pp. 49-53, Jan. 1948.

[110] W. F. Tinney and C. E. Hart, "Power flow solution by Newton's method," *IEEE Transactions on Power Apparatus and Systems*, vol. PAS-86, no.11, pp.1449-1460, Nov. 1967.

[111] R. Tonkoski, L.A.C. Lopes, and T.H.M. El-Fouly, "Coordinated active power curtailment of grid connected PV inverters for overvoltage prevention," *IEEE Transactions on Sustainable Energy*, vol. 2, no. 2, pp. 139-147, Apr. 2011.

[112] S. C. Tripathy, G.D. Prasad, O.P. Malik, and G.S. Hope, "Load-flow solutions for ill-conditioned power systems by a newton-like method," *IEEE Transactions on Power Apparatus and Systems*, vol. PAS-101, no.10, pp. 3648-3657, Oct. 1982.

[113] J. E. Van Ness and H. J. Griffin, "Elimination methods for load-flow studies," *Transactions of the American Institute of Electrical Engineers, Power Apparatus and Systems, Part III.*, vol. 80, no. 3, pp. 299-302, Apr. 1961.

[114] VDE-Study: Energy storage in power supply systems with a high share of renewable energy sources. pp. 1-60, Dec. 2008.

[115] F. A. Viawan, "Voltage control and voltage stability of power distribution systems in the presence of distributed generation," Ph.D. dissertation, Dept. of Energy and Environment. Chalmers University of Technology, Göteborg, Sweden, 2008.

[116] L. H Walker, "10-MW GTO converter for battery peaking service," *IEEE Transactions on Industry Applications*, vol. 26, no. 1, pp. 63-72, Jan./Feb. 1990.

[117] M. Walker, "The improvement of power factor in alternating-current systems," *Journal of the Institution of Electrical Engineers*, vol. 42, no. 195, pp. 599-616, Jun. 1909.

[118] J. B. Ward and H. W. Hale, "Digital computer solution of power-flow problems," *Power Apparatus and Systems, Part III. Transactions of the American Institute of Electrical Engineers*, vol. 75, no. 3, pp. 398-404, Jan. 1956.

[119] M. J. N. V. Werven and M. J. J. Scheepers, "The changing role of energy suppliers and distribution system operators in the deployment of distributed generation in liberalised electricity markets," DISPOWER report, ECN-C--05-058, Jun. 2005.

[120] G. Westinghouse, "Electrical converter." U.S. Patent 366,362, issued Jul. 12, 1887.

[121] H. L. Willis, *Power Distribution Planning Reference Book*, Rev. and Exp. 2nd ed. NY: Marcel Dekker, 2004.

[122] H. Yoshida, "A particle swarm optimization for reactive power and voltage control considering voltage security assessment," *IEEE Trans. Power Syst.*, vol. 15, no. 4, pp. 1232-1239, Nov. 2000.

[123] R. D. Zimmerman, C. E. Murillo-Sánchez, and R. J. Thomas, "MATPOWER: steady-state operations, planning, and analysis tools for power systems research and education," *IEEE Transactions on Power Systems*, vol. 26, no. 1, pp. 12-19, Feb. 2011.

[124] Time-of-Use Prices. (2012)
[Online]. Available: http://www.ieso.ca/imoweb/siteshared/tou_rates.asp?sid=ic/

[125] ENTSO-E Overview of transmission tariffs in Europe. (2012)
[Online]. Available: https://www.entsoe.eu/market/transmission-tariffs/

[126] Reactive Power and Grid Integration with SUNNY MINI CENTRAL and SUNNY TRIPOWER. (2013) [Online].Available: http://www.sma.de/fileadmin/content/global/Solutions/Documents/Medium_Power_Solutions/ReactivePower-UEN101310.pdf

[127] [Online]. Available: http://www.entsoe.eu/(2013)

[128] [Online]. Available: http://www.ieeeghn.org/(2013)

[129] [Online]. Available: http://www.caiso.com/(2013)

[130] [Online]. Available: http://www.iso-ne.com/(2013)

[131] [Online]. Available: http://www.mathworks.com/(2013)

[132] [Online]. Available: http://www.gams.com/(2013)

i want morebooks!

Buy your books fast and straightforward online - at one of world's fastest growing online book stores! Environmentally sound due to Print-on-Demand technologies.

Buy your books online at
www.get-morebooks.com

Kaufen Sie Ihre Bücher schnell und unkompliziert online – auf einer der am schnellsten wachsenden Buchhandelsplattformen weltweit! Dank Print-On-Demand umwelt- und ressourcenschonend produziert.

Bücher schneller online kaufen
www.morebooks.de

 VDM Verlagsservicegesellschaft mbH
Heinrich-Böcking-Str. 6-8　　Telefon: +49 681 3720 174　　info@vdm-vsg.de
D - 66121 Saarbrücken　　　Telefax: +49 681 3720 1749　　www.vdm-vsg.de

Printed by Books on Demand GmbH, Norderstedt / Germany